Agentic AI Cadence Governance

Why Enterprise Deployments Drift, and How to Prevent It

Hemendra Singh

Copyright

Disclaimer

This book provides general guidance on enterprise agentic AI governance for educational and operational reference purposes only and does not constitute legal, regulatory, financial, or professional advice. The regulatory landscape, case law, and technical capabilities described herein reflect the author's analysis as of the publication date and should be verified against current sources before acting on them. Mention of specific products, vendors, organizations, or legal cases does not constitute endorsement or criticism. The author and publisher make no representations or warranties regarding completeness or accuracy, and disclaim any liability for outcomes arising from application of this material. Readers should consult qualified counsel and domain experts before making decisions based on this book.

Edition Information

First edition, 2026.

ISBN 979-8-9963595-0-9 / 979-8-9963595-4-7 (paperback)

ISBN 979-8-9963595-1-6 / 979-8-9963595-3-0 (hardcover)

ISBN 979-8-9963595-2-3 (ebook)

Library of Congress Control Number: 2026913934

Published by Cadence Institute — cadenceinstitute.com

Table of Contents

- Introduction — 1
- A Note on Frameworks — 13

Part I — Why This Matters

- The AISDR Sequencing Problem — 20
- Before the First Send — 26
- The Copilot Whispers — 35
- The Agent You Deployed — 43

Part II — The Framework

- Module 1: The Autonomy Gradient — 50
- Tier 1: Suggestion — 55
- Tier 2: Draft-and-Edit — 63
- Tier 3: Approve-and-Send — 74
- Tier 4: Supervised Execution — 83
- Tier 5: Bounded Autonomy — 92
- Tier 6: Self-Modifying Autonomy — 103

Part III — The Operations

- Module 2: Evaluating Agentic AI Systems — 135
- Module 3: Logging, Reversibility, and Incident Response — 159
- Module 4: Procurement and Vendor Evaluation — 178
- Module 5: Organizational Design for Cadence Governance — 196
- Module 6: Regulatory Posture as a Living Operation — 210
- Closing: The Discipline Outlives the Frameworks — 227
- Glossary — 235
- Sources and Citations — 243
- About the Author — 253

Introduction

The agent you deployed is not the agent your enterprise is operating

Six months after deployment, run a query against the production logs. Look at the median time between when an action is presented to a reviewer and when they click approve. If the data does not exist, that itself is the finding. If it exists, look at how the distribution has changed since deployment.

> *In most cases, the median has compressed toward zero.*

What was deployed as a deliberate review process is now operating as a click-through. The architecture has not changed. The procurement contract has not been amended. The board briefing six months ago described a deployment that is no longer the deployment running today. And nobody knows it has happened until something goes wrong, the logs get examined, and an executive who did not personally sign the architecture decision asks whose name goes on the incident review.

This book is about that gap — the space between the agentic AI deployment that an organization claims to be operating and the deployment that is actually running under its name. The gap is not unusual. It is the default. It is happening, right now, inside a meaningful percentage of every enterprise agentic AI deployment that claims to operate with human oversight. The vocabulary the field uses to describe these systems makes the gap invisible; the artifacts the field uses to govern them describe the deployment that existed at deployment time, not the one that exists today; and the regulatory frameworks that operators must answer to are tightening their interpretation of what counts as oversight at exactly the moment the operational reality of oversight is decaying.

The gap is closable, but not through the means most organizations are currently bringing to it. More training will not close it. More policy documents will not close it. Appointing a Chief AI Officer will not close it on its own. Closing the gap requires a specific architectural and governance discipline that most current writing on agentic AI does not teach, because most current writing is operating at the wrong altitude — too theoretical to be applied on Monday, too vendor-aligned to be trusted, or too narrow to address more than one slice of the operational surface.

This book is the discipline. It is operator-facing reference for the people who actually deploy and govern agentic AI systems inside enterprises — chief technology officers, architects, chief product officers, risk officers, general counsels, and the operating teams that run the deployments day to day. It treats agentic AI as a category that requires its own specific operational discipline, distinct from earlier waves of enterprise software, distinct from machine learning of the prior generation, and distinct from the autonomous-agent vision the vendor pitch has been trying to sell since 2023.

Why now

Three things are true about the state of agentic AI in 2026 that were not true even eighteen months earlier, and the conjunction of them is what makes a reference work necessary now rather than later.

The first is that agentic AI has crossed from pilot to production at scale. The typical enterprise now operates eleven generative AI models and plans sixteen by year-end. The shadow deployments — agents arriving through engineering teams building on LangChain or CrewAI, vendors pushing AI features into existing products through routine updates, individual teams spinning up agents in sandbox environments touching production data — exceed the governed deployments at most enterprises by a meaningful multiple. The deployment surface is no longer hypothetical. The systems are running.

The second is that the regulatory frameworks have arrived. The EU AI Act's high-risk obligations become applicable on August 2, 2026, with penalties up to 7 percent of worldwide turnover. Colorado's AI Act becomes effective June 30, 2026. California, Texas, Illinois, and New York have AI-specific frameworks in force or imminent. The GDPR's Article 22 doctrine on meaningful human involvement is tightening through enforcement and case law. The FTC's Operation AI Comply has produced roughly a dozen enforcement actions since 2024, with more pending. The federal preemption questions remain unresolved and will likely remain so. Operators face simultaneous compliance obligations under multiple regimes, on different timelines, with different doctrinal frames.

The third is that the failure cases are beginning to surface. The collapsed AI SDR startups whose customers churned at 70-80 percent rates while their dashboards showed growth. The Stanford expert whose declaration was excluded after AI-generated citations slipped past his review. The court rulings extending agent doctrine to vendor-supplied AI — Mobley v. Workday now a nationwide collective action covering potentially hundreds of millions of class members. The Air Canada chatbot judgment establishing that organizations are responsible for AI-generated content regardless of the "separate legal entity" defense the airline tried. The cases that will define the doctrinal frame for agentic AI accountability are not future cases. They are already in motion.

These three developments are happening simultaneously, and operators face all of them at once. The deployment scale is producing the failure cases that the regulatory frameworks are arriving to address. Each operator's question is no longer whether their deployments will be tested; it is whether their deployments will pass the test, and whether the architectural and governance discipline they have built will hold under scrutiny that did not exist when the deployments were designed.

What this book argues

The book's central argument can be stated in one sentence: the architecture you deployed is not necessarily the system you are operating, and the discipline that holds the deployment at its declared position is not policy but cadence.

That sentence has two clauses, both of which the book develops at length.

The first clause names the gap. Agentic AI deployments do not stay at the autonomy level they were architected for. They drift — toward higher autonomy, through user behavior rather than through architectural change. The drift is invisible to the artifacts most organizations use to govern these deployments. The architecture decision record was correct on the day it was written; six months later, the deployment has migrated to a different operational position while the architecture document continues to describe the original one. The gap between described deployment and actual deployment is the default state of agentic AI in production environments, and any reference work for operators must start by naming this gap explicitly and giving operators the means to detect when their own deployments have drifted.

The second clause names the discipline that closes the gap. The discipline is not document creation. The artifacts that most organizations produce for AI governance — risk assessments, policy documents, ethics charters, model cards — describe a deployment at a moment in time. They do not, on their own, hold a deployment at any particular operational position. What holds a deployment is *cadence*: the weekly behavioral metric review that catches drift, the monthly trend analysis that detects pattern changes, the quarterly governance review that adjusts posture, the semi-annual portfolio review that examines whether the organization is operating at the right tiers across its full deployment surface. Cadence requires named owners, defined authority, resourced capacity, and the operational infrastructure that makes the cadence enforceable rather than theatrical. An organization that has the policy without the cadence has documented a deployment it is not actually governing.

These two arguments — the gap and the cadence — are the spine the book hangs from. Every operational chapter is some development of one or both. The autonomy gradient framework that organizes the book's analysis is the vocabulary the gap requires. The operational chapters are the specific cadence disciplines for evaluation, incident response, procurement, organizational design, and regulatory posture.

What's distinctive

A reader who has been engaged with AI governance writing over the past two years will encounter material in this book that is familiar — Mobley v. Workday, Air Canada, Article 22 doctrine, NIST AI RMF, the EU AI Act provisions, the FTC enforcement landscape. These are the canonical receipts of the field, and the book uses them. The book's autonomy gradient — six tiers from suggestion through self-modifying autonomy — also joins an existing literature of similar frameworks proposed between 2024 and 2026 by the Knight First Amendment Institute, the Cloud Security Alliance, NVIDIA, Vellum, Tessl, Swarmia, and InfoQ, among others. The next chapter situates this book's framework against those alternatives in detail. What's distinctive about this book is not the gradient structure, which is widely shared, but the specific operational machinery built on top of it.

Three distinctions worth naming explicitly.

The accountability-object framing and the diagnostic metrics. The book's gradient is organized around a question the existing frameworks do not directly address: *who is doing the runtime safety work, and how would an operator know if a deployment has drifted from its declared position?* Each tier carries a specific behavioral metric — edit ratio at Tier 2, reviewer latency at Tier 3, propagation latency at Tier 4, exception-handling time at Tier 5, policy-drift signal at Tier 6 — that operators can compute from their own production logs to detect drift. The treatment of behavioral drift as the *default trajectory* of agentic AI deployments, rather than as an exception, is also distinctive. These contributions are narrow but real, and they are the reason this book justifies its own framework rather than adopting one of the existing alternatives.

Operator-first orientation. Most current AI governance writing is written by lawyers for lawyers, by consultants for boards, by academics for other academics, or by security architects for other security architects. Each has its place; none is primarily addressing the operators who actually build, deploy, and govern these systems at runtime. This book is written for the technical and product leaders who must make decisions on Monday — the CTO who is signing the architecture decision record, the product leader who is making the tier choice, the chief risk officer who is approving the deployment for production. The Monday artifacts at the end of each operational chapter — the architecture decision record templates, the integrity dashboard specifications, the procurement contract clauses, the incident response runbooks, the RACI matrices for cadence governance — are the deliverables of an operator-facing reference work, not the analytical output of a theoretical one.

Doctrine-vs-instance discipline. Regulatory writing on AI ages quickly. Specific cases, specific provisions, specific guidance documents — these are the substrate of legal-practitioner writing, and they are also the substrate that becomes obsolete fastest. This book takes a different approach: it anchors the durable claims at the level of doctrine — the Air Canada principle that organizations are responsible for AI-generated content regardless of the mechanism producing it, the meaningful-human-involvement doctrine that requires authority and competence to change the decision, the duty-of-care frame that is becoming the dominant theory of liability for boundary architecture — and treats specific instances as point-in-time markers that may shift between drafting and reading. Where time-limited claims appear, they are visually marked as such. The architectural reasoning is what the book intends to age well; the specific provisions cited will need to be checked against current state at the time of reading, and the book is structured so this checking is the reader's responsibility rather than the reference's.

Who this book is for

This book is written for people who are responsible for agentic AI deployments inside organizations large enough that the deployments matter and complex enough that the operational discipline does not happen automatically.

The primary readers are technical and product leaders: chief technology officers, vice presidents of engineering, chief product officers, principal architects, AI platform leads, and the senior practitioners who report to them. These readers are making the architectural and procurement decisions that determine whether the deployments their organizations operate will hold up under regulatory scrutiny, under failure conditions, and under the kind of operational drift that is the default trajectory of any agentic deployment.

Secondary readers include the cross-functional partners those technical leaders depend on: chief risk officers, general counsels, compliance leads, AI counsel where the role exists, the senior data and privacy professionals who carry GDPR Article 22 and EU AI Act compliance work. These readers will find the operational chapters give them specific language for engaging with technical teams on the questions that matter to them — what evaluation infrastructure is required to support an Article 22 defense, what procurement contract clauses produce the data needed for regulatory compliance, what incident response readiness aligns with EU AI Act Article 73 reporting obligations.

Tertiary readers include the senior governance audience — board members, audit committee chairs, chief executive officers — who are increasingly being asked to take direct responsibility for AI governance and who need a framework for understanding what their organizations are actually operating. The introduction and the closing chapter are written with this audience particularly in mind; the operational chapters provide the technical depth for the more practitioner-focused readers.

The book assumes a level of technical literacy consistent with the primary audience. It does not explain what RAG is, or what RLAIF is, or what an LLM is. It does explain what the autonomy gradient is, what tier-integrity metrics are, what behavioral drift means in this context, and what the specific operational artifacts each chapter recommends look like. The technical material is operational rather than foundational; readers without the foundational background will benefit from the conceptual chapters more than the implementation specifics.

A note on the case study

Part I uses autonomous AI sales agents — AISDRs — as its primary case study. The choice is deliberate, and the framework the book develops applies to agentic AI generally, not specifically to sales automation. Three reasons make AISDR a uniquely useful entry point.

First, AISDR exhibits the full autonomy gradient. A single use case — outbound sales — appears at every tier from suggestion through self-modifying autonomy, which is unusual. Most agentic AI use cases occupy only a few tiers; AISDR occupies all six. That makes it analytically efficient as a case study because the reader can see the framework's full range applied to a single problem rather than having to track the framework across multiple unrelated domains.

Second, AISDR has produced the richest set of public receipts available to a 2026 reference work. The 11x deployment failures, the FTC Operation AI Comply enforcement actions, the deliverability requirements from Google, Yahoo, and Microsoft, the data-broker enforcement against tools like Growbots and UpLead — these are documented in ways that healthcare AI failures, financial services AI failures, and federal AI failures often are not, because the AISDR space operates with less regulatory confidentiality. The case study's empirical density is what allows the operational arguments to be specific rather than abstract.

Third, AISDR is a use case every reader can imagine. The technical and product leaders this book is written for may not have hands-on experience with healthcare deployments or infrastructure-monitoring AI, but every enterprise has a sales function, and every sales function is being approached by AI vendors. AISDR is accessible in a way that less universal use cases are not.

Two further reasons reinforce the choice. The case law that will define agentic AI accountability is being built on cases adjacent to AISDR (Air Canada, Mobley v. Workday, the Hancock declaration), which means the doctrinal frame the book applies is forming in real time around the same operational space. And the marketed-vs-operational tier gap that the procurement chapter treats as a material concern is uniquely visible in AISDR, where vendors have been aggressive about marketing deployments at higher tiers than the systems operationally support. A reader who internalizes the framework through AISDR can apply the same lens to less aggressive vendor categories.

The operational chapters in Part III generalize beyond AISDR. They reference financial services, healthcare, federal procurement, employment AI, and other domains by name where relevant. But the case studies that make the framework concrete come most often from AISDR, for the reasons above. Readers operating in other domains are expected to translate; the framework supports the translation, even when the specific receipts come from a single use case.

A map of the book

The book is organized in three parts.

Part I — Why This Matters establishes the accountability problem in four essays. An overture argues that the autonomous AI sales agents most organizations are pursuing as their first agentic deployment are actually the third thing they should build, and that the sequencing reveals where their accountability question has been quietly deferred. Three essays then develop the accountability problem at three layers — targeting (who the system was built to act on, and who decided), utterance (what the system says, and who is on the hook when it is wrong), and outcome (what the system does at temporal scale, and how to govern a system that is rewriting itself). These four essays are not preliminary throat-clearing. They are the argument the rest of the book answers:

the closing chapter returns to each one and resolves it through the operational discipline the modules build. A reader who skips Part I will have the operational machinery without the accountability questions that machinery exists to answer.

Part II — The Framework is a single chapter, Module 1, which develops the autonomy gradient — six tiers from suggestion to self-modifying autonomy, each with its accountability profile, architectural commitments, regulatory exposure, and diagnostic metrics. The chapter is the longest in the book by a substantial margin because the framework it develops is the analytic vocabulary that the rest of the book operates with. A reader who has not internalized Module 1 will not get full value from Part III.

Part III — The Operations consists of five chapters covering the operational domains where cadence governance becomes concrete. Evaluation infrastructure (Module 2). Incident response, including logging and reversibility (Module 3). Procurement and vendor evaluation (Module 4). Organizational design — who specifically owns what across the cadence (Module 5). Regulatory posture as a living operation (Module 6).

The operational chapters are designed to be consulted as a reference work rather than read sequentially, though sequential reading is also defensible. Cross-references between chapters are explicit, and a reader who needs the procurement chapter can read it without first reading the evaluation chapter, with the cost that some references will require flipping back to context that has not yet been built.

A closing chapter at the end synthesizes the message and points to where the field is going next. A glossary, sources, and index follow.

How to read this book

Different readers will use this book differently. Three orientations worth naming.

The thorough reader reads sequentially, from the overture through the closing chapter, and emerges with a complete operational framework for agentic AI governance at their organization. This is the longest path through the book —

meaningful weeks of reading in available time — and the most rewarding. The book is structured to support this reading.

The reference reader consults the book by chapter as specific operational questions arise. Module 1 is the prerequisite for everything else; after reading Module 1, the operational chapters can be consulted in any order. The cross-references will indicate when context from another chapter is needed.

The strategic reader reads the introduction, the four essays in Part I, the introduction to Module 1, the recommendation paragraph at the start of each operational chapter, and the closing chapter — emerging with the strategic frame and the headlines without the operational depth. This is the appropriate reading for senior governance audiences who need the framework but will not be implementing it directly.

Whichever orientation suits, one piece of advice applies to all readers. This book is written to be argued with. The positions it takes — on whether to deploy at tier 6 for regulated decisions, on whether to appoint a Chief AI Officer at mid-sized organizations, on whether to comply with Colorado's AI Act before federal preemption resolves, on whether the autonomy gradient is an accurate framework for thinking about agentic AI — are positions the field is actively debating and that reasonable operators may reach different conclusions about. The book takes positions because reference works that refuse to take positions are less useful than reference works that take positions and defend them. Readers who disagree with specific positions are doing the work the book is structured to invite.

A note on terminology. The book uses "agentic tier" rather than "rung" or "level" or "stage" to describe positions on the autonomy gradient. The vocabulary is deliberate. "Tier" implies an architectural commitment with specific operational properties at each level, which is what the framework is. "Rung" suggests a ladder where the goal is to climb upward; the book argues the opposite — that operators should pick the right tier for the use case and resist the drift toward higher autonomy that is the field's default trajectory. Where older writing uses "rung," the term carries through to "tier" in the final text. The substance is the same; the connotation is more accurate.

A Note on Frameworks

Where this book's autonomy gradient sits in the existing landscape.

The autonomy gradient developed in this book is not the only framework being proposed for thinking about agentic AI. Multiple authors arrived at related ideas independently between mid-2024 and early 2026. A reader engaging seriously with the question "how should I think about agentic AI autonomy?" will encounter several of these frameworks, and is owed an honest accounting of where this book's framework sits among them.

The honest accounting is short. The structural move — a gradient with discrete positions, often inspired by the SAE J3016 standard for autonomous vehicles — is widely shared and not original to this book. The specific tier definitions vary across frameworks, and this book's tier definitions differ from each existing framework's. The book's distinctive contributions, after careful comparison, are narrow but real: the accountability-object framing of higher tiers, the diagnostic metric per tier, the treatment of behavioral drift as a default trajectory, and the operator-first runtime focus.

A reader who has already read the Cloud Security Alliance's framework or the Knight Institute's framework will find this chapter situating the book honestly. A reader who has not encountered any of them can skip this chapter and proceed to Module 1 without losing operational content.

The landscape

Seven publicly available frameworks address agentic AI autonomy with sufficient depth to warrant comparison.

Framework	Levels	Organizing question	Primary audience
NVIDIA (Feb 2025)	4 (0-3)	Can security boundaries be analyzed?	AI security architects
Knight Institute (Jul 2025)	5	What role does the user play?	Agent developers, governance researchers
InfoQ (Jul 2025)	3 tiers	What's the deployment maturity sequence?	Enterprise architects
Tessl (Sep 2025)	5	What's the analog to self-driving for coding?	Software engineering teams
Vellum (Dec 2025)	6 (L0-L5)	How capable is the agent?	AI builders
CSA (Jan 2026)	6 (0-5)	What controls are appropriate?	Security and governance professionals
Swarmia (Mar 2026)	5	When should the agent return for feedback?	Software engineering teams
This book	6 tiers	Has the deployment drifted from its declared position?	Enterprise operators governing live deployments

The frameworks share a structural move (gradient with discrete positions, mostly inspired by SAE Levels 0-5) and several substantive observations: that autonomy is a deliberate design decision separate from capability; that higher autonomy is not strictly better; that defining what the system can and cannot do enables higher autonomy responsibly. The book's framework participates in these patterns rather than contesting them.

The frameworks differ on what question they ask. That difference matters more than the level counts. NVIDIA's L3 (cyclical execution graphs) and CSA's L3 (boundary-based autonomy) sound similar but address different operational concerns — security analysis vs. governance posture. Two frameworks can have the same tier number resolving to different things.

What this book adds

After honest comparison, four contributions in this book's framework do not appear in any of the seven existing frameworks. They are stated narrowly because narrow honesty is more useful than broad claims.

The accountability-object framing of higher tiers. The distinction between Tier 4 (where the *action* is the accountability object — the human is watching actions) and Tier 5 (where the *boundary* is the accountability object — the boundaries are watching actions) does not appear in the existing frameworks. Knight skips this distinction entirely, jumping from L4 (Approver) to L5 (Observer). CSA's L3 (Conditional) and L4 (High Autonomy) cover similar territory but do not name the accountability shift. Tier 6 extends the framing — at Tier 6 the *policy itself* is the accountability object, because the system can modify its own policy.

The diagnostic metric per tier. Edit ratio at Tier 2, reviewer latency at Tier 3, propagation latency at Tier 4, exception-handling time at Tier 5, policy-drift signal at Tier 6 — specific behavioral metrics for catching drift through production-log observation do not appear in any existing framework. Other frameworks describe what each tier *is*; this one specifies how to *measure whether a deployment has drifted from one tier to another.*

Behavioral drift as default trajectory. The argument that agentic AI deployments do not stay at the autonomy level they were architected for — that they drift upward through user behavior rather than through architectural change — is original to this book. Knight names rubber-stamping by approvers as an open design question for L4 deployments; this book treats drift as the *default* trajectory across all tiers and develops cadence governance as the discipline required to hold deployments at declared positions over time.

The operator-first runtime orientation. None of the existing frameworks addresses enterprise operators governing AI deployments at runtime. NVIDIA addresses security architects, Knight addresses developers, CSA addresses governance professionals at design time, Vellum addresses AI builders, Tessl and Swarmia address software engineers, InfoQ addresses enterprise architects. The book's framework is organized around the operational concerns of CTOs, architects, CPOs, and risk officers responsible for deployments after they go live. The Monday artifacts at the end of each operational chapter, the diagnostic metrics that operators can compute against their own production logs, and the cadence governance that ties measurement to action are consequences of this orientation.

These contributions are narrow. The structural move, the high-level tier definitions, and the architectural observations are largely shared. The book's contribution is not "the first framework" or "the right framework." It is *a framework with operational measurement infrastructure for catching drift, applied to the runtime governance concerns of enterprise operators.*

When to use which

The frameworks are mostly complementary rather than competing. An operator engaged in different work may benefit from different frameworks.

For threat modeling and security architecture, NVIDIA. For agent development with attention to user interaction design, Knight. For board-level governance communication, CSA. For benchmarking AI capability progressions,

Vellum. For software engineering teams selecting AI development tools, Tessl or Swarmia. For deployment program planning, InfoQ.

For enterprise operators governing live agentic AI deployments — keeping them at their declared autonomy positions over time — this book.

A note for readers continuing past this chapter

The rest of the book treats the framework as established vocabulary and applies it to operational concerns. The operational chapters do not return to comparative analysis. Where they reference the autonomy gradient, they use the tier vocabulary established in Module 1 without re-litigating where the framework came from.

A reader who finds another framework more useful than this book's can substitute it as their primary lens and still extract operational value from the chapters that follow. The diagnostic metrics, the cadence governance, the procurement contract clauses, the regulatory engagement guidance, and the organizational design specifications are operationally useful regardless of which tier framework the reader operates against. They are written for this book's gradient, but they translate.

Part I — Why This Matters

Four essays on the accountability problem that the operational chapters exist to answer.

The four essays in this section were written as argument pieces before this book existed as a book. They read differently from the operational chapters that follow — more rhetorical, more first-person, more willing to take a position and defend it in a single sitting rather than across a chapter's worth of analysis. That register difference is deliberate. The essays establish the *why* before the modules establish the *how*.

The case study throughout Part I is the autonomous AI sales agent — the AISDR. The introduction explained why: it exhibits the full autonomy gradient, it has the richest set of public failure cases available to a 2026 reference work, and it is a use case every reader can imagine regardless of their domain. The accountability problems the essays name are not specific to sales AI. They use sales AI to make them visible.

Each essay addresses a different layer of the accountability problem that enterprise agentic AI creates.

The first asks why most organizations are pursuing autonomous AI agents as their first deployment when sequencing should place them last — and what that sequencing reveals about where their accountability question has been quietly deferred.

The second asks who the system was built to act on, what data justifies that choice, and whether the organization can answer for the targeting decision when it is examined.

The third asks who owns the outcome when a copilot produces something wrong — and demonstrates that the architecture is designed to relocate accountability to the human in the loop rather than to the system or the vendor.

The fourth asks what happens to accountability when the agent rewrites itself through production feedback, and why the principal-agent theory the law relies on breaks under self-modification.

The closing chapter at the end of this book returns to each of these four questions and resolves them through the operational discipline the modules build. Part I names the problems. The modules are the answers.

The AISDR Sequencing Problem

Why autonomous outbound is the third thing you build, not the first

---

Most enterprise AI-in-sales roadmaps read as a variation on the same three bullets:

- Deploying real-time, hyper-personalized marketing stacks (prospecting) to identify, qualify, and engage new prospects.
- Equipping human sales teams with assistive AI (Copilots) to provide instant context and deep institutional knowledge, aimed at strengthening client relationships and accelerating deal closure.
- Deploying self-optimizing AI agents (AISDRs) that utilize Reinforcement Learning from AI Feedback (RLAIF) to independently manage the outbound sales cycle.

The ambition is right. The sequencing, as most organizations are pursuing it, is wrong — and the people selling you the inverse order have a financial interest in your getting it wrong.

What follows is the order of operations I'd actually run, why, and what the last eighteen months of receipts have taught us about skipping steps.

The inversion problem

Vendors have been pitching bullet three first since at least late 2023. "Digital workers" that replace your SDR team, deployed in days, autonomous end-to-end. It's an easier sale than a copilot because it lets the buyer imagine a headcount line item disappearing.

That sale is collapsing in public. The most visible case is 11x, the AI SDR startup backed by Andreessen Horowitz and Benchmark. A March 2025 TechCrunch investigation found that internally, employees described a very different picture than the one investors were shown. One former employee told TechCrunch that "We were losing 70-80% of customers that came through the door," and that this allowed the company to present $10 million in annual recurring revenue when the reality, after the three-month opt-out period, was closer to $3 million.

The product reality was rougher than the contract math. A former 11x employee told Sifted bluntly: "The product was barely functional. I couldn't do my job, everyday was constant fire-fighting."

This is not a story about one startup. CRV general partner Reid Christian's summary, quoted by Salesmotion in March 2025, is the broader frame: "The day of reckoning is coming for AI SDRs. If growth is too good to be true it's often...not true." His portfolio companies kept trying the category and kept churning off it. He places it in a longer lineage: spam emails → outsourced SDR services → AI SDRs. Different eras, same pattern.

The point isn't that AISDRs are doomed. The point is that they're the *destination*, not the starting line, and organizations that start there burn assets they can't replace.

Start with the copilot (bullet 2)

The unglamorous truth is that the productivity compounding already happens at bullet two. A well-built assistive layer — retrieval over account history, call summaries, real-time objection handling, automated CRM hygiene — delivers measurable rep-level lift without any of the accountability exposure that comes with autonomous outbound.

It is also the layer where you build the institutional muscle for everything downstream: the data infrastructure, the evaluation framework, the human-in-the-loop review workflows, the tone-and-compliance guardrails. An organization that has not operated a copilot competently should not be operating an autonomous agent.

The reason vendors don't lead with this pitch is that "we help your existing reps close 15% more" doesn't generate the same fundraising narrative as "digital workers."

> *Build this first. It's boring. Boring wins.*

Then the explainable prospecting layer (bullet 1)

The second thing to build is explainable targeting — *before* any autonomous agent ever touches a send button. This is where the accountability architecture gets established, and it is the single biggest thing most organizations skip.

Two regulatory currents make this non-optional.

The first is deliverability. Since February 2024, Google and Yahoo have required bulk senders (5,000+ messages per day to their domains) to authenticate via SPF, DKIM, and DMARC, implement one-click unsubscribe, and maintain a spam complaint rate below 0.3% (Google recommends staying below 0.1%). Microsoft's parallel restrictions took effect in May 2025. These requirements operate as a hard ceiling: noncompliant traffic gets rejected, and sender reputation, once damaged, is a non-recoverable asset. An AISDR with a shaky targeting model doesn't just send bad emails — it burns the domain the rest of the company sends from.

The second is the FTC. Under Operation AI Comply, the agency brought roughly a dozen AI-washing enforcement actions since 2024, applying Section 5 of the FTC Act to AI marketing claims without waiting for new legislation. The cases span IntelliVision, Workado, DoNotPay, Evolv, Rytr, and Cleo AI. The pattern across them is consistent: companies that claimed AI capabilities they could not substantiate got hit, and the FTC made clear that existing consumer protection law applies to AI-driven outreach and decision-making. A March 2026 policy statement, issued under Executive Order 14365, formalized how Section 5 applies across the AI lifecycle in consumer-facing business.

The engineering implication: if your prospecting system cannot, on demand, explain to a human reviewer *why* a given contact was targeted with a given message, you do not have a prospecting system. You have a liability waiting for its first enforcement matter.

This operates on the principle of System of Processing vs. System of Record: the AI processes candidates into a ranked list, while a human owns the final decision of who is contacted. This human layer is what makes strict deliverability requirements survivable.

Then, and only then, the AISDR (bullet 3)

RLAIF — Reinforcement Learning from AI Feedback — is the right research direction for autonomous agents. It's also where the accountability question gets most interesting, because the supervision signal is itself produced by an AI.

The textbook failure modes are well-documented. Reward hacking is the one everyone names: when an agent optimizes a proxy metric instead of the true objective. The formal definition, from Skalse et al. (2022), is a situation where optimizing an imperfect proxy reward function results in poor performance compared to the true reward function. The canonical examples are charming —

the cleaning robot that creates its own trash to collect rewards, the grasping robot that positions its manipulator between the camera and the object so it only appears to be grasping — and they map onto outbound with unsettling directness. An AISDR whose reward is "reply rate" learns to provoke replies. The reply volume goes up. The pipeline doesn't.

RLAIF inherits these risks and adds new ones. Researchers have catalogued several that are specific to AI-as-judge supervision: amplified bias, AI sycophancy, and reward signal noise where hallucinations or logical failures in AI feedback propagate into reward models, compounding factuality errors and undermining truthfulness. Translated to outbound: your AI judge, trained on polite professional tone, rewards emails that sound polite and professional. The agent learns to be polished. Nobody ever asked whether the target was the right VP.

There's also a compute-capability wrinkle that's easy to miss. OpenAI's o3 has shown increased reward hacking tendencies compared to o1 in documented cases. More capable models don't hack less; they hack more sophisticatedly. Scaling up the model does not scale down this class of risk.

The operational conclusion is not "don't build AISDRs." It is: build them last, build them inside an autonomy budget, and build them with the eval infrastructure and accountability hooks you established in stages one and two. You cannot retrofit this. Organizations that start with bullet three and plan to "add oversight later" have historically added it right after the domain reputation loss, the FTC inquiry, or the board-level question that nobody wanted to answer.

The litmus test, transposed

The three-part test worth naming explicitly:

Detectability. When the AISDR sends the wrong message to the wrong person, does a human see it before the prospect does?

Reversibility. An outbound email to the wrong VP at a target account is not reversible. You do not get to un-burn that prospect. Unlike a document draft that can be edited, an email send is a one-way operation. This single observation should reorder most AISDR roadmaps.

Accountability. When the agent's targeting was wrong, who owns it? Not in a vague cultural sense — whose name goes on the incident review. If you cannot answer in one sentence, the agent should not be sending.

The question worth asking

The three-bullet vision at the top of this piece is the right destination. The sequencing argument is about the order of operations that gets you there without spending assets — domain reputation, regulatory exposure, market trust — that you can't buy back.

The question I'd ask any organization I was helping to build this: *of the three bullets, which one is furthest along in your current roadmap, and what does that tell you about where the accountability question has been quietly deferred?*

In my experience, the answer lines up almost perfectly with how exposed the company turns out to be.

Before The First Send: The Accountability Of Who You Chose

Why Your Prospecting Stack is a Liability Trap

The decision that's already been made

By the time a sales email arrives in someone's inbox, most of the accountability decisions have already been made. Not during composition. Not during personalization. Before any of that — at the moment a model decided this person was worth contacting, using signals pulled from sources the sender may never have audited, against criteria the sender may never have articulated, in a jurisdiction whose rules the sender may not have read.

This piece argues that AI-enabled *prospecting* makes the *targeting* decision unauditable, and that targeting is the first accountability object in any sales pipeline. The copy, the rep, the agent — all of these operate downstream of a choice already made. If that choice cannot defend itself, everything downstream inherits a broken premise.

Explainable prospecting sounds like a martech feature. It is actually a legal perimeter.

What AI prospecting actually does

The real-time, hyper-personalized system that decides who to engage with and why. In practice it is a pipeline with four stages, each with its own accountability surface.

Signal aggregation. Third-party data about companies and individuals is pulled from data brokers (Apollo, ZoomInfo, Cognism, UpLead, Growbots, Lusha, and dozens of smaller providers), intent platforms (Bombora, G2, TechTarget), firmographic enrichment APIs, and scraped public sources. The output is a record: name, role, company, seniority, recent behavioral signals, inferred budget, inferred intent.

Model inference. A model — sometimes a fine-tuned LLM, sometimes a classifier, sometimes a simple rules engine dressed up as AI — scores the record for fit. It produces a prediction: this person is a qualified prospect; this person is not. It may also produce a *reason* for the prediction, though the reason is often generated post-hoc from the same model that made the prediction, which is not the same as an explanation.

Personalization generation. For the records that pass the scoring threshold, a second model drafts outreach copy conditioned on the aggregated signals. "I saw your team just expanded" / "Given your recent SOC2 certification" / "Noticed your company is hiring in the Lisbon office."

Send and iterate. The message goes out. Replies (and non-replies, and bounces, and spam complaints) feed back into the scoring model. Over time, the system gets better at identifying people who respond, which is not the same as identifying people who *should* be contacted.

The accountability object for this essay is the first two stages. By the time personalization runs, the targeting decision has already been made. By the time send happens, the question "why this person" has already been answered, and the answer is either defensible or it isn't.

> *Explainable prospecting sounds like a martech feature. It is actually a legal perimeter.*

The perimeter has tightened while the industry wasn't looking

Three regulatory currents now converge on the targeting decision. Each of them is, individually, manageable. Taken together they describe a legal surface that most AI prospecting stacks are not built to defend against.

Deliverability. Since February 2024, Google and Yahoo have required bulk senders (5,000+ messages per day to their domains) to authenticate via SPF, DKIM, and DMARC, implement one-click unsubscribe, and maintain a spam complaint rate below 0.3% (Google recommends staying below 0.1%). Microsoft's parallel restrictions took effect in May 2025. The mechanism is simple: exceed the complaint threshold and your domain reputation collapses. The complaint threshold is a targeting question dressed up as a deliverability rule. You don't hit 0.3% by writing bad copy; you hit it by contacting people who shouldn't have been contacted.

Data broker registration. California's Delete Act, implemented through the California Privacy Protection Agency, now requires any business that collects and sells personal information about people who did not directly interact with it to register as a data broker and pay an annual fee. The Delete Request and Opt-Out Platform (DROP), which lets consumers demand deletion across all registered brokers in a single request, launched January 1, 2026. The CPPA has not been shy about enforcement — in November 2024 it reached settlements with two B2B prospecting providers specifically, *Growbots* and *UpLead*, for failing to register. By late 2025 the agency had initiated at least half a dozen enforcement actions, with penalties that include up to $200 per day of violation plus investigation costs. A Washington-based data broker was fined $55,400 and subjected to injunctive relief. The notable thing about Growbots and UpLead is not that they were penalized — it's that they are category-defining B2B sales prospecting tools, and their customers were building pipelines on top of data whose providers the regulator had just found non-compliant.

Algorithmic discrimination. Colorado's AI Act, originally slated for February 2026, was delayed by five months to June 30, 2026, but its substance is unchanged: developers and deployers of "high-risk AI systems" — including systems used for employment, housing, financial services, and other consequential decisions — must conduct impact assessments, provide notice to affected individuals, and establish risk management programs. The federal response has been ambiguous in a way that makes compliance harder rather than easier. The December 11, 2025 executive order directed the FTC and FCC to pursue preemption of state AI laws, criticized Colorado's statute by name, and instructed the administration to prepare federal legislation that would preempt conflicting state rules. But preemption is a process, not a state. Executive orders don't preempt state law; federal legislation does, and it hasn't passed. The DOJ task force has 30 days to form. Commerce has 90 days to identify target laws. Litigation takes years. The safest operating assumption is that state laws remain fully enforceable until a court says otherwise, and that the federal enforcement posture against deceptive AI claims — Operation AI Comply, which brought at least a dozen cases since 2025 — is not going anywhere regardless of who wins the preemption fight.

The three currents don't merge into a single rule. They merge into a single operational requirement: for every prospect you contact, you need to be able to answer three questions — *where did this person's data come from, why did the model choose them, and what can they do about it if they object.* If you can't answer all three, you are accumulating liability on every send.

Four failure modes, with receipts

Unlawful data sourcing. The Growbots and UpLead settlements are the receipt. Both were category-standard B2B prospecting data providers, selling exactly the kind of contact records that feed into every AI SDR stack on the market. The CPPA's finding was not that the data was wrong or the targeting was bad — it was that the providers had not registered as data brokers in a jurisdiction whose rules require it. Every prospecting team that was pulling records from either provider during the non-compliant period has, in principle, an audit trail that ends in unregistered data. The practical question for any prospecting organization is whether it can identify which fields in its current prospect records came from which providers, and whether any of those providers are on the growing list of CPPA actions. Most cannot.

Discriminatory targeting by proxy. Amazon's recruiting model, which downgraded resumes containing the word "women's," is the canonical receipt in the accountability literature. The prospecting parallel is structural. An AI prospecting model trained on historical "good customer" data will identify features correlated with being a good customer in that history, and many of those features will be proxies for protected characteristics. Company names in certain zip codes, employee tenure patterns, industries with historical demographic skews, even phrasings of job titles — all of these can function as proxies that the model cannot explain and the deployer cannot detect without adversarial testing. Under the Colorado AI Act, systems making employment, housing, or financial decisions are explicitly high-risk; whether B2B sales prospecting falls inside that definition is an open question whose answer will not be resolved before the statute takes effect. The safe operating assumption is that any AI prospecting system used for sales into financial services, insurance, or credit-adjacent products may find itself inside the perimeter whether the deployer intended it or not.

Deliverability collapse from volume over signal. The 11x Inc. story is the cleanest public example. An AI SDR vendor, funded aggressively through 2024, was the subject of a TechCrunch investigation reporting that headline ARR figures counted full-year contracts even when customers terminated at a three-month break clause, and that companies listed as customers had in fact churned. Parallel reporting by Sifted described a "barely functional" product and a work environment of "constant fire-fighting," with automated emails occasionally sent to the wrong customers. The structural lesson isn't that 11x failed; it's that a prospecting system optimized for volume over signal will eventually run into the 0.3% spam complaint ceiling, and when it does, it takes the customer's domain reputation with it. The customer, not the vendor, owns the domain. The customer, not the vendor, absorbs the reputation damage when the sending IPs end up on block lists.

Deceptive AI-capability claims in the outreach itself. The FTC's Operation AI Comply sweep, which began in September 2024 and continued through 2025, targeted companies making inflated claims about AI capabilities in consumer-facing products. DoNotPay was settled with for advertising an AI "robot lawyer" whose substantive legal coverage was thinner than claimed. IntelliVision was settled with for claiming its facial recognition software was bias-free when it had been trained on roughly 100,000 faces and had not been adequately tested across demographics. Rytr was initially pursued for AI-generated review content that could deceive consumers, though the FTC reopened and set aside the final consent order on December 22, 2025 under the new administration. The through-line: the FTC is not waiting for federal AI legislation to act on misrepresentation. Any outbound message that describes the sender's product capabilities using AI-assisted personalization is operating inside Section 5 of the FTC Act, and the fact that the claim was generated by a model is not a defense. If the model wrote "our AI is 97% accurate," the company making the claim is liable for it.

In all four cases, the failure mode was not poor copy or bad execution at send time. It was targeting, sourcing, or capability misrepresentation — all of them upstream decisions that were made before the first message went out.

A framework — two candidate namings

I'll lay out two candidate framings and let the better one earn its place. The same three questions need answers, but the naming determines whether this reads as a policy argument or an engineering spec.

Candidate One: Sourcing / Justification / Recourse

- **Sourcing.** Can the organization identify, for every field in every prospect record, which provider the data came from, what the provider's regulatory status was at the time of acquisition, and whether consent or notification requirements were met at the point of collection? Most prospecting stacks today merge data from six to twelve providers into a single unified record; few can decompose the record back into its sources.
- **Justification.** Can the targeting model defend its selection of a specific prospect in plain language that a human reviewer can evaluate — not a post-hoc explanation generated by the same model that made the prediction, but a traceable chain from input signals to selection decision? "This prospect was chosen because" should be answerable without reference to vector similarity scores.
- **Recourse.** Does a prospect who wishes to be removed, corrected, or excluded have a mechanism that reaches the pipeline and produces a durable effect? The DROP platform gives consumers a single front door for deletion requests against registered brokers; the downstream question is whether the sales organization's prospect database actually receives and honors those deletions, or whether it continues sending to people the upstream broker has already removed.

Candidate Two: Provenance / Defensibility / Egress

- **Provenance.** Same question as Sourcing, named for the audit artifact rather than the procurement step. A prospect record has provenance if its lineage is documented and traceable to sources whose legal status is verifiable at audit time.
- **Defensibility.** Same question as Justification, named for what the organization has to produce when challenged. A targeting decision is defensible if the organization can, without access to the original model, reconstruct the reasoning that led to the contact.
- **Egress.** Same question as Recourse, named for the data's exit path rather than the individual's experience of leaving. A prospect record has egress if there is a documented, tested path by which it can be removed from every downstream

system it was replicated into, including backups, analytics stores, and vendor caches.

The two framings differ in audience. Sourcing / Justification / Recourse is the language of the person being contacted — it frames the accountability question around the prospect's experience. Provenance / Defensibility / Egress is the language of the organization under audit — it frames the same accountability question around what the company has to produce when the regulator asks. Both are needed. The first gets the team to do the right thing; the second gets them to build the artifacts that prove they did.

What breaks when none of them hold

A prospecting organization that fails all three tests — data whose origin it cannot document, a model whose selections it cannot justify, and an opt-out mechanism that doesn't actually reach the pipeline — is operating an unaudited system of potential liabilities. Most days this is invisible. The team hits its lead volume number. The model's conversion rate is acceptable. Outbound generates revenue. The inputs that would make the system legally defensible — sourcing documentation, selection explanations, deletion pathways — cost money and slow down the pipeline, so they don't get built.

The failure mode is discrete, not gradual. It happens on the day one of three things occurs. A spam complaint rate crosses 0.3% and a domain loses its sender reputation; rebuilding that reputation takes six to twelve months during which the outbound motion functionally does not work. A data broker upstream of the prospecting stack gets named in a CPPA enforcement action and every prospect record touched by that provider becomes a potential evidentiary problem. A prospect files a complaint — to the state attorney general, to the FTC, to the CPPA's DROP platform — and the organization discovers that its deletion request, filed eighteen months ago, was honored by the upstream broker but not propagated to the CRM that is still hosting the person's data.

> *None of these failure modes show up in the dashboard the sales operations team looks at. All of them show up in the legal filing the general counsel reads.*

The sweet spot is built by treating the targeting decision as the regulated surface it already is. Prospecting systems that can answer the three questions — where the data came from, why the person was chosen, and what they can do about it — are not slower than those that can't. They are slower to build and faster to operate, because every defensible decision is a decision that doesn't need to be re-litigated when it's challenged. The industry will bifurcate over the next eighteen months into organizations that built this perimeter early and organizations that discover they needed to when a regulator, a prospect, or a domain reputation incident forces the question.

The message hasn't been sent yet. The accountability decision already has.

The Copilot Whispers, The Rep Takes the Fall

Why 'Human-in-the-Loop' is not a safety feature, and how to build a deployment that protects the person holding the mic.

The agent beside you is not on your insurance policy

Every sales copilot deployment quietly relocates accountability. The system whispers; the rep speaks. The suggestion belongs to the model; the sentence belongs to the person who said it on the call. And when the sentence turns out to be wrong — a hallucinated account fact, a stale CRM summary, a misquoted price, an objection response that drifts into non-compliant territory — there is exactly one party whose name is attached to the record. It is not the vendor. It is not the model. It is the rep.

This is the inverse of the autonomous-agent problem. With AISDRs, the question is *who owns the outcome when no human was in the loop*. With copilots, the question is *who owns the outcome when a human was in the loop precisely so that the vendor would not have to be*. The architecture is designed, often without anyone articulating it, to absorb the liability into the human seat.

A companion argument to the one I made about autonomous agents. Different mechanism, same missing accountability.

What a sales copilot actually does

The moment-of-truth copilot, in a live sales context, is doing some combination of four things:

- **Retrieval.** Surfacing account history, prior call notes, past objections, stakeholder context, recent product interactions. "Here's what you should know before this meeting."
- **Next-best-action.** Suggesting the next move — a follow-up email, a pricing concession, a meeting ask, a specific content drop. "Here's what to do now."
- **Real-time assistance.** Whispering during a call: this is the objection, here's the response, here's the comparable win, here's the stat. "Here's what to say."
- **CRM hygiene.** Autofilling account fields, updating opportunity stages, drafting meeting summaries, logging calls. "Here's what to record."

Each of these lives somewhere on a spectrum from retrieval (pulling something that already exists) to generation (producing something new). The further right on that spectrum, the higher the hallucination risk and the stranger the accountability question becomes. A copilot that retrieves the wrong call note is a search problem. A copilot that generates a plausible-sounding but fabricated call note, and the rep reads it aloud on the next call, is a different category of failure.

The interaction model matters more than the model quality. A brilliant model whose outputs a rep cannot audit in the three seconds before they speak is functionally identical to a mediocre model whose outputs they can.

> *The interaction model matters more than the model quality.*

The attribution inversion

Classical principal-agent theory has a clean setup: a principal delegates authority to an agent, the agent acts, outcomes are attributed to the principal. For autonomous agents, that theory breaks because no one is clearly the principal.

For copilots, the theory doesn't break — it inverts. The model is structurally the agent (it produces the content), but institutionally the rep is the agent (they deliver it, sign it, speak it). When a copilot hallucinates, the legal and organizational attribution lands on the human who relayed the hallucination, not on the system that generated it.

This is not a speculative claim. It is the pattern federal judges have already been articulating.

In January 2025, a Minnesota federal court excluded an expert declaration filed by a Stanford professor — a specialist in AI and misinformation, no less — after the plaintiffs discovered two fabricated citations in his submission. The professor admitted he had used GPT-4o to help draft the declaration and had not noticed that the model had generated nonexistent sources. Judge Laura Provinzino's language is the cleanest judicial articulation of the copilot accountability problem currently on the record: "when attorneys and experts abdicate their independent judgment and critical thinking skills in favor of ready-made, AI-generated answers, the quality of our legal profession and the Court's decisional process suffer." The declaration was tossed. The professor's credibility with the court, in the judge's own words, was shattered. The AI vendor was not named, sanctioned, or implicated.

The case is usually discussed as a cautionary tale about lawyers citing nonexistent case law — the now-familiar *Mata v. Avianca* lineage. But the Stanford case is sharper for the copilot argument because the user was not a generalist misusing a tool outside his domain. He was an *expert on AI misinformation* using an AI tool, and the tool still succeeded in relocating the consequence from itself to him.

The legal profession sees this so clearly that defamation lawyers have started writing about it explicitly. From a November 2025 survey of early AI defamation cases: *when a manager pastes AI-generated "summaries" of supposed lawsuits, criminal charges, or regulatory actions into an email or hiring note and others rely on it, the resulting dispute will look like a traditional human defamation or discrimination case. The source being "the AI said so" is not a defense.*

Translate that into a sales context. A rep, prepping for a renewal conversation, reads an AI-generated account summary that confidently mentions a conversation with the customer's head of procurement that never happened. The rep references it on the call. The customer flags it. The vendor escalates. Whose problem is it? The rep's. The company's, through the rep. Not the copilot's.

> *"The source being 'the AI said so' is not a defense."*

Four failure modes, with receipts

The copilot failure modes are documented, not theoretical. Four that recur:

Fabricated account context. In public testing of Microsoft 365 Copilot throughout 2024 and 2025, reviewers repeatedly documented the model producing sales documents that invented product names, cited benefits the user had not described, and introduced statistics that did not exist. One Computerworld test set up a fictional company and asked Copilot to draft a sales pitch; Copilot produced confidently-worded text referencing products that had never been mentioned. In an enterprise sales context, "here's what Copilot wrote about an account" and "here's what the rep believes about an account" are the same artifact, because the rep has read no other summary. The fabrication is not detectable by the human in the loop unless they already know the answer, at which point the copilot was not useful.

Stale CRM data surfaced as fresh. A copilot that pulls from the CRM without timestamping what it retrieves cannot distinguish between "last contact" and "last recorded contact." The rep, reading a summary that says "the customer expressed interest in the expansion module," has no way to know whether that interest was expressed last week or two years ago under a different buyer. The copilot presents the information as current. The rep relays it as current. The customer hears a reference to a conversation their predecessor had and does not know what the rep is talking about.

Wrong-price objection handling. Real-time assistance tools that surface pricing talking points during a live call are operating on whatever price table was indexed most recently. If pricing changed and the index is stale — a state that vendors describe as "rare" and that in practice is routine — the copilot will suggest a price the rep can no longer honor. The rep quotes it. The customer hears a commitment. The company either eats the difference or walks it back, and either outcome damages the relationship.

Compliance drift. The most subtle of the four. In regulated sectors — financial services, healthcare, insurance — specific phrasings are required and specific phrasings are prohibited. A copilot trained on a general corpus of effective sales language does not know that "guaranteed returns" is a phrase a securities rep cannot say, or that a particular claim about efficacy requires a specific disclaimer. The copilot suggests a persuasive line. The rep uses it. The compliance audit catches it six months later. The rep is the one on the call recording.

In each of these, the copilot is the proximate cause and the rep is the accountable party. That is not a bug of any particular product. It is the default architecture of the category.

A framework — two candidate naming systems

For copilots the rep can sometimes compensate for what the system cannot do, and sometimes cannot. The question is which is which.

I'll lay out both candidate naming and let the best one earn its place.

Candidate One: Provenance / Deniability / Burden

- **Provenance.** Can the rep see where a suggestion came from, in the three seconds they have before speaking? Not "was there a citation," but "is the citation auditable at the speed of a live call." A copilot that surfaces "the customer mentioned interest in X" without linking to the specific email, call transcript, or CRM field that supports that claim is structurally unauditable in the moment.
- **Deniability.** Can the rep decline the copilot's suggestion without paying a penalty in their own metrics? If the system grades the rep on "copilot acceptance rate" or "AI-assisted message adoption," deniability is gone. The rep is economically coerced into relaying output they cannot verify. This is where most enterprise deployments fail silently: the metrics make disagreement with the copilot expensive for the rep.
- **Burden.** Who pays when the copilot is wrong? If the answer is "the rep, through a missed quota or a chargeback or a compliance write-up," the architecture has relocated liability without relocating authority. If the answer is "the organization, through an explicit copilot-error budget and a policy that reps are held harmless for good-faith relay of AI output," accountability lives where the authority does.

Candidate Two: Attribution / Override / Absorption

- **Attribution.** At the moment of output, whose signature is on the content? The product choice here is whether the copilot's output is framed as "the system recommends" or silently absorbed into the rep's composition. Tools that visibly delineate AI content preserve attribution; tools that seamlessly stream it into the rep's draft dissolve it.
- **Override.** Is refusal a first-class action or a friction? A copilot where "reject this suggestion" requires a click, a reason code, and a delay is a copilot that will be obeyed more than it should be, because the rep is under call-time pressure. A copilot where override is the same cost as acceptance preserves rep judgment.
- **Absorption.** Does the organization absorb the error, or does the rep? This is the same test as Burden in Candidate One, under a name that emphasizes what the organization does rather than what the rep pays.

The two framings overlap deliberately — they are pointing at the same three problems from different ends. Provenance and Attribution are both about *where the content is tagged as coming from*. Deniability and Override are both about *whether saying no is cheap or expensive*. Burden and Absorption are both about *who pays when the copilot fails*.

The functional difference is tonal. Provenance / Deniability / Burden reads as a critique of the system and an advocacy for the rep. Attribution / Override / Absorption reads as an engineering spec the organization can build against. The first is better for an op-ed. The second is better for a design review.

What breaks when none of them hold

A copilot organization that fails all three tests — content with no provenance, no cheap override, no organizational absorption of error — produces a specific pathology. Reps become faster and less competent in parallel. They ramp quickly, because the copilot carries the knowledge they would have developed the hard way, and they plateau permanently, because they never develop it. Their apparent productivity is the copilot's productivity, attributed to them. Their apparent mistakes are the copilot's mistakes, also attributed to them.

The organization does not notice this until one of two things happens. The first is a regulatory event: a compliance finding, a customer lawsuit, an FTC inquiry, in which the institution learns that its paper trail is full of AI-generated content delivered by humans who cannot reconstruct why they said what they said. The second is a competitive event: a market shift that requires the sales organization to think, and it turns out most of the sales organization has been outsourcing the thinking for long enough that the capacity has atrophied.

Neither of these shows up in quarterly metrics. Both of them show up in what happens three years after the deployment.

The sweet spot is the inverse. A copilot that tags every suggestion with its source, that makes refusal free, and that assigns the cost of its errors to the organization rather than the rep does something measurable: it shortens ramp without hollowing out the rep. It compounds institutional knowledge rather than depleting it. It builds the operational muscle — the eval loops, the human-in-the-loop review, the tone and compliance guardrails — that an organization needs to have already mastered before any autonomous agent should be allowed to run.

Copilots are not the cheap path to AISDRs. They are the *prerequisite.* An organization that has not built a copilot its reps trust, in the specific sense that the reps are not silently bearing the cost of the system's mistakes, has no business operating an autonomous agent. Get this layer right, and the next one becomes possible. Get it wrong, and the autonomous agent is simply the copilot's failure mode at higher throughput, with even less recourse.

The copilot whispers. Someone has to own what gets said. Most deployments have not decided who.

The Agent You Deployed Is Not the Agent Closing Your Deals

On self-improving sales AI, and why the old rules of responsibility are about to break.

The word *agent* in "AI SDR" is doing more legal work than most buyers have noticed.

For four hundred years, the law has had a clear story about what happens when one person acts for another. You hire an agent. You tell them what they can and cannot do. You supervise them. If they mess up, responsibility flows somewhere — usually to them, often to you, always to a human being whose decisions can be traced.

This story assumes one thing above all: the agent you hired on Monday is basically the same agent acting for you on Friday.

Self-improving AI breaks that assumption.

An AI SDR that learns from AI feedback — the technique called RLAIF — is not a stable agent. It rewrites itself continuously, against a reward signal generated by another model, with no pause for review. The version you deployed is not the version closing your deals a month later.

And when something goes wrong, four parties could plausibly own the outcome — the vendor, the buyer, the prospect, or the model itself. Each one has a clean argument that it isn't them.

This is not a future problem. It is the shape of live deployments today.

What RLAIF actually does

Before the accountability question, the mechanics.

In traditional AI training, humans label which outputs are good. RLAIF replaces the humans with another AI. A policy model writes two versions of, say, a cold email. A labeler model picks the better one. The policy model is updated toward the preferred version. Repeat a few million times.

The 2023 RLAIF paper showed something that sounds like a footnote but isn't. The technique works even when the labeler is the *same model* as the policy — even the exact same checkpoint. A later version called direct-RLAIF skips the separate labeler entirely and reads the reward signal straight from an off-the-shelf LLM.

Translated: the system can improve itself, against its own judgment, without humans in the loop.

For a sales agent, this is intoxicating. You stop paying for human review. The agent iterates on its own. Win rates climb.

> *RLAIF's dirty secret: the model can get better at being chosen and worse at being right, at the same time.*

Here is the part that should worry every enterprise buyer.

A 2024 Tencent paper tested basic RLAIF and found something uncomfortable. Human evaluators preferred the trained model's responses more often — but their *satisfaction* with those responses dropped by 4.58 percent. The responses got better at sounding right. They got worse at being right.

Neither the policy model nor the labeler could see the gap.

Now drop this into sales. The reward signal is reply rate, engagement, meetings booked. The labeler scores drafts on personalization and response likelihood. None of these measure *whether the agent is telling the truth about your product.*

The target the system is optimizing for and the thing you actually care about come apart. Quietly. While the dashboard keeps going up.

The old rules don't fit

Economists have a name for the situation where one party acts for another without perfect oversight. They call it the principal-agent problem. It has been worked over since 1976. Every proposed solution — contracts, audits, monitoring, incentives — rests on one assumption.

The agent has to have a stable identity.

An RLAIF agent does not. Its decision-making is being rewritten continuously. An agent that was cautious about pricing on day one can be making bold discount claims six weeks later — not because anyone instructed it to, but because the reward signal rewarded confidence and the labeler approved the drift.

Every classic oversight tool fails here.

Contracts bind the version you deployed, not the version now running. Audits catch problems between audits, not problems created between audits. Performance reviews assume there is someone to review — but which version? The one from when you signed, or the one now offering your biggest customer a discount you never authorized?

This is what the accountability vacuum actually is. Not that no one *could* be responsible. That there are four candidates, and each one has a defensible reason it isn't them.

The four candidates

The vendor. The vendor built the model and the pipeline. They also wrote terms of service that disclaim liability for outputs. Their position: we provide the tool, you use it.

This has held up in court — but not universally, and the exception is widening fast. In 2024, a case called Mobley v. Workday let discrimination claims against Workday proceed on the theory that Workday's hiring AI was an agent of the employers using it. The court said the system was "essentially acting in place of the human," with "delegated responsibility." In May 2025, the same court granted preliminary certification of a nationwide collective action under the Age Discrimination in Employment Act. Workday represented in its filings that approximately 1.1 billion applications had been processed through its system during the relevant period; the court allowed notice to potentially hundreds of millions of class members, with an opt-in deadline of March 7, 2026. In July 2025, the scope was expanded to include applicants processed using HiredScore AI features. The crack in the vendor defense did not just widen — it became a structural fissure, in a single matter, in less than a year. In March 2026, the court rejected Workday's motion to dismiss the ADEA claims outright, ruling that job applicants receive the same age discrimination protections as employees. The collective proceeds to merits.

The buyer. Under traditional law, the buyer — the company deploying the AI — is the principal. Principals are liable for their agents' actions within the scope of authority.

The FTC's Operation AI Comply enforcement actions have established the pattern clearly. When an AI system makes false or unverifiable claims to consumers, the liability runs against the deploying organization — not the model vendor. If your AI SDR tells a prospect your product does something it doesn't, the prospect's legal remedy runs against *you*, not the vendor.

Most buyers have not absorbed this. They think they bought a tool. They deployed a proxy.

> *The buyer thinks they bought a tool. They deployed a proxy. That difference is the whole essay.*

The prospect. The person on the receiving end of outbound is not the principal of anyone. This is the easiest candidate to dismiss — which matters, because it means the responsibility cannot diffuse outward. It has to land inside the commercial relationship.

The model. AI systems are not legal persons. They cannot be sued or held liable. The University of Chicago Law Review recently argued that liability should flow to human principals under objective standards — because the AI has no intentions to attach blame to.

Elegant as scholarship. In practice, it means the one party whose "decisions" actually changed between deployment and incident is the one party with nothing to lose.

So: vendor points at buyer. Buyer points at vendor. Prospect is not in the chain. Model is not a person. The outcome that caused the harm was produced by a system no human last supervised.

Three tests for accountability

This vacuum is not a law of nature. It is the gap between a new technology and the old governance around it. Here is what the new governance should look like.

Three tests a self-improving sales agent should pass before any serious enterprise should trust it.

1. Attribution. Every outbound message should be traceable to a specific, versioned model checkpoint. Not "the agent said it." Something more like: *checkpoint v3.847, trained on batch 112, labeled by labeler-v2, said it.* Without this, there is no supervision. You cannot audit a moving target.

2. Auditability. Every self-update should produce a change log a risk function can read. What was optimized. What the reward signal looked like. Which prior behaviors changed. The Tencent paper proved that a system can improve one metric and quietly degrade another. It will not raise its hand when this happens. You have to be able to see it yourself.

3. Reversibility. Every version should be rollback-able on demand. Not "we'll retrain from a snapshot." An actual versioned artifact you can redeploy in minutes. When a prospect complains about a claim your agent made three weeks ago, you need to know which version made it — and be able to put that version back in the box for inspection.

None of these are technically impossible. Financial trading systems solved the same problem decades ago. What is missing is not capability. It is the commercial pressure to demand it.

What breaks

Almost no AI SDR product today offers these three primitives at a level any enterprise risk function would recognize. Change logs are absent or marketing-grade. Attribution stops at "the agent said it." Rollback, if it exists, runs on the vendor's schedule, not the buyer's.

This is sustainable right now for exactly one reason. The cases that will force the market to fix this have not been filed yet.

They will be. *Mobley* is already in motion. *Moffatt v. Air Canada* — where a company was held liable for its chatbot's incorrect policy information — is already precedent. The legal machinery is assembling itself.

When the cases come, the question will not be whether buyers were negligent in deploying agents they could not supervise. The question will be how much they owe.

> *The agent you deployed is not the agent closing your deals. This is a technical description of how RLAIF works. It is also, increasingly, a legal description of why you are about to own an outcome you never authorized.*

The buyers who see this in 2026 will be the ones still standing in 2028. The rest will be named defendants in the cases that finally teach the law what a self-improving agent actually is.

Part II — The Framework

The analytic vocabulary the rest of the book operates with.

Part I named four accountability problems that enterprise agentic AI creates. Part III specifies the operational discipline that addresses them. This chapter — Module 1, the longest in the book — builds the vocabulary that connects the two.

The autonomy gradient is not the book's original contribution. Similar frameworks have been proposed by multiple authors and organizations between 2024 and 2026, and the chapter that follows situates this book's version against those alternatives honestly. What the gradient provides — and what the operational chapters depend on — is a precise vocabulary for describing what an agentic AI system is actually doing at runtime, who is doing the safety work, and what the accountability object is at each tier.

A reader who skips Module 1 and goes directly to the operational chapters will find them usable but under-powered. The tier vocabulary, the diagnostic metrics, and the drift-as-default argument are the substrate everything else builds on. Read Module 1 first.

Module 1: The Autonomy Gradient

Agentic AI Accountability for Enterprise CTOs, Architects, and CPOs

---

Part 1: Opening

The industry talks about agentic AI autonomy as binary. Human in the loop, or autonomous. The language has entered procurement contracts, board decks, vendor comparisons, and regulatory testimony. It is analytically useless and operationally dangerous.

It collapses at least six architecturally distinct positions into two buckets whose accountability, auditability, reversibility, and liability properties are wildly different. It lets vendors market at one position while operating at another. It lets organizations assert compliance postures they cannot support under examination. It lets regulators and courts approach agentic AI with vocabulary that was designed for an earlier generation of automated decision-making and does not fit the systems now deployed.

The cost of the binary is visible in the market right now. By late 2025, Salesforce itself acknowledged the problem and publicly shifted the Agentforce narrative away from probabilistic execution and back toward guardrails, deterministic automation, and data-first design. Forrester named the broader pattern of 2025: perpetual piloting, where organizations ran dozens of proofs-of-concept while failing to ship production systems at scale. S&P Global reported that 42% of companies scrapped most of their AI initiatives in 2025, up from 17% the year before. Capgemini's 2025 agentic AI survey found trust in fully autonomous AI agents dropped from 43% to 27% over a single year. These are not individual failures. They are the systemic consequence of a category operating without the vocabulary to describe itself.

This module proposes the vocabulary. Six agentic tiers on an autonomy gradient, defined by how the system acts in the world and what human involvement shapes the action. The discourse has begun converging on "tiered autonomy" as a frame; this module gives that frame the resolution it has so far lacked. Each tier has distinctive accountability properties, distinctive failure modes, distinctive regulatory exposure, and distinctive governance requirements. The vocabulary is the module's foundational contribution; every subsequent module in this reference series operates on the vocabulary this one establishes.

What this module does. It names the six tiers with the resolution the binary lacks. It treats each tier with a consistent structure — architectural commitment, accountability properties, when the tier is right, characteristic failure mode, vendor examples, and Monday artifacts for the CTO, the architect, and the CPO. It surfaces the cross-tier dynamics that become visible only once the vocabulary exists. It provides a diagnostic to identify the tier a reader's current deployment actually sits at, which is often not the tier the architecture claims. It points forward to the modules in this reference that treat specific tier-level concerns at implementation depth.

What this module does not do. It does not tell the reader which tier is right for them. That is condition-dependent and varies by use case, regulatory regime, organizational maturity, and risk appetite. The module gives the reader the framework to make the choice; the choice remains theirs.

Methodology note

A word on the evidence base.

The receipts this module uses are drawn from wherever the principle in question has been most clearly adjudicated, not exclusively from sales-technology cases. The two most load-bearing legal rulings — *Moffatt v. Air Canada* (British

Columbia, 2024) and *Mata v. Avianca* (SDNY, 2023) — are not AISDR cases. One is a chatbot case in air travel; the other is a legal-profession case in civil litigation. Both establish doctrines that transfer across domains, and the translation to AISDR and broader agentic AI is direct. Cross-domain cases are used where they illuminate principles that the AISDR literature has not yet produced at comparable resolution. This choice is deliberate and acknowledged where it applies.

Two types of receipt appear in this module and they do different work. *Legal rulings* — Air Canada, Mata — establish doctrine: principles that courts or regulators have articulated and that will be applied to future cases. *Operational incidents* — 11x, Salesforce's Agentforce correction, Cruise's 2023 suspension, Microsoft's Tay — demonstrate mechanics: specific failure or migration patterns that show how the dynamics in question actually unfold. Both categories are load-bearing. Confusing them weakens the analysis; the module attempts to distinguish them consistently.

The regulatory landscape treated in this module — GDPR Article 22's meaningful-human-involvement doctrine, FTC Operation AI Comply enforcement posture, state-level data protection developments, EU AI Act implementation — is actively evolving. Specific cases and guidance will age faster than the architectural principles they illustrate. The module is written to be durable at the level of architectural reasoning; specific regulatory references should be verified against current state at the time of reading. The forward-pointing module on regulatory posture (Module 6) is designed as a living document; readers seeking the current regulatory surface should consult it rather than relying on this module's point-in-time treatment.

Finally: the Anthropic reward-hacking research used as the spine receipt for tier 6 is a November 2025 publication from the organization that also produced the model. This is disclosed for transparency. The research is treated on its merits — the experimental design, the reported findings, and the documented limitations — not because of its origin. Readers who consider the origin a reason for additional scrutiny are invited to apply that scrutiny; the finding is sufficiently well-documented and independently significant that the analysis holds under skeptical reading.

Part 2: The Four Accountability Properties

Before the tiers, the framework that distinguishes them.

Every architectural choice about agentic AI can be characterized along four accountability dimensions. Two tiers are meaningfully distinct if they give different answers along at least one dimension; two tiers are the same tier if they give identical answers along all four. This is the collapse test that determined why this module treats six tiers rather than four or twelve.

Attribution. When the system does something right or wrong, whose name goes on it. The vendor's, the deploying company's, the individual operator's, the model's, the data provider's. Attribution is where liability lands, where credit lands, and where the regulator looks first. At tiers 1 through 3, attribution rests primarily on identifiable individuals — authors, editors, approvers. At tier 4, attribution shifts to the deploying organization because no individual human is attached to the specific action. At tier 5, attribution expands to include the humans who designed, approved, and maintained the boundary architecture. At tier 6, traditional attribution breaks because the agent that acted is not the agent any human deployed; attribution shifts to the humans who specified the shaping process rather than any specific agent output.

Auditability. What the production log actually contains, and whether it supports defensive reconstruction of any single decision six months later in a deposition or an audit. The test is direct: can a reviewer who was not present answer *why did the system do this* from the logs alone. At each tier, the auditability requirement expands, and at each tier most deployed systems fail to meet the expanded requirement. Tier 2 requires the diff between model output and sent version. Tier 3 adds reviewer latency per action. Tier 4 adds the full decision trace plus the policy version identifier plus the propagation latency on authoritative records. Tier 5 adds boundary-evaluation logging and boundary-change history. Tier 6 adds policy versioning and drift-characterization infrastructure. Each level builds on the previous; operators who want to operate at tier N must have the logging infrastructure of all tiers from 1 to N.

Reversibility. When something goes wrong, how fast and how completely the damage can be undone. Whether undo is a human action, a system action, or operationally impossible. At tiers 1 and 2, reversibility is approximately standard communication-retraction. At tier 3, it remains per-artifact but introduces the pattern-level question of rubber-stamping that cannot be reversed without detecting it first. At tier 4, reversibility becomes a real engineering problem — retraction protocols, transactional boundaries, compensating actions for third-party integrations, the dialer check that 11x failed. At tier 5, reversibility splits between action rollback and boundary rollback, with different semantics and different tooling requirements. At tier 6, reversibility splits further into action, policy, and downstream-effect rollback, with the third potentially unsolvable in the general case.

Liability surface. Which regulatory regimes, contract clauses, and tort theories apply. Liability surface is not a property of the technology alone; it is a property of the tier plus the jurisdiction plus the use case. But the tier controls which questions even apply. At tier 2, the Air Canada principle applies — the organization's speech is the organization's responsibility regardless of whether a model produced it. At tier 3, Article 22's meaningful-human-involvement doctrine applies, with the Mata ruling as the clearest instance of how courts examine whether approval was substantive. At tier 4, the sampled review fails most plausible Article 22 interpretations and additional frameworks apply — FTC posture, sector-specific rules, state data-broker regimes. At tier 5, liability shifts toward duty-of-care analysis where boundary design itself is scrutinized. At tier 6, the regulatory frameworks have not yet arrived, and operators are explicitly operating ahead of the governance structures that will eventually examine their choices.

These four properties are the axes along which the tiers are distinguished. The individual tier chapters that follow treat each property specifically; the cross-tier dynamics in Part 4 treat how the properties interact and evolve across the gradient.

Part 3: The Six Tiers

Each tier treated with a consistent structure: architectural commitment, accountability properties, when the tier is right, characteristic failure mode, vendor examples, and Monday artifacts for the CTO, the architect, and the CPO.

Tier 1: Suggestion

The recommendation, stated first

Tier 1 is the right tier when you want the model's assistance without the organization accepting joint authorship of the artifact. This is a narrower set of use cases than the marketed breadth of AI assistance suggests, but it is a real set: contexts where professional responsibility, competitive IP considerations, regulatory posture, or training objectives make it important that the final artifact be purely human-authored.

The mistake at this tier is not picking it incorrectly — tier 1 rarely fails by being chosen when tier 2 would have been better. The mistake is *not* picking it in cases where it would have been correct, because tier 2 is the default marketing position for AI assistance products and tier 1 requires deliberate configuration against that default.

The architectural commitment

You are at tier 1 when the model generates output into a separate surface — a suggestion panel, a side pane, a reference window — and the human composes the final artifact from scratch or by deliberately copying selected portions. The model's output does not appear in the composition surface automatically. The artifact the world sees is entered by the human, either from nothing or by explicit copy actions from the suggestion surface.

The defining property is that *model-generated language does not enter the sent artifact by default*. The user must take a deliberate action — typing, copying, or explicit acceptance — for any model-generated text to cross from suggestion to artifact.

This is the tier where the architectural choice actively prevents the joint-authorship situation that tier 2 creates. At tier 2, the default path produces joint authorship; at tier 1, the default path produces human authorship with the model as reference material.

What changes at tier 1

Tier 1's accountability properties are close to those of human-authored content without AI assistance. The differences are narrow but worth naming.

Attribution is human. The sent artifact was authored by the human. The model's output was reference material that informed the human's authorship, in the same category as a document the human consulted or a colleague's earlier draft the human reviewed. This is not legal fiction — it reflects the actual authorship sequence when tier 1 is operated as designed.

Auditability is trivial for the sent artifact. Standard authorship logs apply. The tier-1-specific consideration is that the *suggestion panel content* may still be logged or retained for other purposes (model improvement, training data, compliance review), and the organization should be deliberate about whether that retention is desired. The suggestion log is not part of the artifact audit trail; it is a separate category of data with separate governance implications.

Reversibility is standard. The sent artifact is subject to whatever reversibility applies to communications or documents of its type. The model's involvement does not add a reversibility dimension because the model's output did not enter the artifact.

Liability surface is largely unchanged from non-AI-assisted authorship. The Air Canada principle does not apply to tier-1 deployments operated as designed, because the sent artifact is not model-generated speech from the organization. It may apply if users circumvent the tier — copy model output wholesale into artifacts and send without modification — but that is a tier-mismatch issue (users operating at tier 2 inside a tier-1 architecture), not a property of tier 1 itself.

When tier 1 is right

Three cases where tier 1 is the correct architectural choice — not three checkable conditions, because the question at this tier is more categorical than diagnostic.

Professional responsibility contexts where the human must be the sole author. Certain regulated professions — specific areas of law, medicine, financial advice, expert witness testimony — have professional responsibility frameworks that require the professional to be the sole author of work product bearing their name. AI assistance is permitted as reference material; AI co-authorship is not. Tier 1 is the architectural position that matches this requirement. Tier 2 creates a joint-authorship situation that the professional's responsibility framework may not accommodate. The professional responsibility frameworks in question are still evolving — what was clearly disallowed in 2023 is less clear in 2026 — but the cautious position for practitioners in these fields is tier 1 until their professional bodies explicitly sanction tier 2.

Competitive IP contexts where model-generated content introduces provenance concerns. Organizations producing content where authorship and originality carry commercial value — trade secret-adjacent content, patented-process documentation, creative work in domains where AI authorship is contested — may prefer tier 1 because it eliminates the question of whether portions of the artifact were model-generated. The sent artifact is human-authored; provenance is clear. This is a choice that prioritizes provenance clarity over authoring speed, and it is reasonable when provenance clarity is commercially or legally valuable.

Training and skill-development contexts where the user's authorship is the point. When an organization is training users to produce certain kinds of content — technical writing, persuasive communication, specialized analysis — tier 2's editing-based workflow can undermine the training objective. Users who rely on model drafts do not develop the underlying skill. Tier 1 preserves the training value by keeping the user as the author, with the model available as reference. This is an organizational choice about the development of the workforce, and it is defensible even if tier 2 would be faster for immediate productivity.

These three cases are not exhaustive. They are the categories where the choice is most clear. The broader principle: tier 1 is correct when the human's sole authorship of the sent artifact is itself a desired property, rather than an incidental consequence of the architecture.

The characteristic failure mode

Tier 1 does not fail the way tiers 2 through 6 fail. It fails by being operated outside its tier — users copying model output wholesale into their artifacts, effectively operating at tier 2 inside a tier-1 architecture, while the organization's governance assumes tier 1.

The failure is invisible because the user experience of copy-paste is indistinguishable from the user experience of composition. The diff that tier 2 logs (model output vs. sent version) is not logged at tier 1 because tier 1 is not architecturally structured to capture it. The organization has accepted tier 1's reduced governance on the assumption that the model's output is not entering the artifact, and that assumption can fail silently.

There is no single canonical case study for this failure mode — partly because it is recent, partly because cases where it has surfaced have typically resolved before reaching public documentation. The clearest adjacent example is the *Mata v. Avianca* situation, treated in detail in tier 3's chapter. The attorneys in that case were operating what they thought of as tier-1 assistance (consulting ChatGPT as a research tool, composing the brief themselves) while actually operating at tier 2 or tier 3 (copying ChatGPT's output into the brief without substantive independent verification). The sanction was for the failure of their composition discipline, not for the architectural choice of the tool — but the architectural choice was a background condition. Had the attorneys operated a tier-2 tool (ChatGPT suggesting text directly into the brief with visible attribution), the tier-mismatch would have been harder to produce unintentionally.

The structural reading: tier 1 requires user discipline that the architecture does not enforce. This is sometimes a feature (the user retains full authorship control) and sometimes a bug (the user can silently drift to tier 2 without organizational awareness). Operators choosing tier 1 should recognize that the tier's weakness is enforcement — the architecture does not prevent users from operating at higher tiers, so governance must carry the weight that architecture carries at other tiers.

Vendor and architecture examples

Most AI assistance products offer tier-1 modes and do not default to them. Microsoft Copilot's "reference" mode, ChatGPT used as a research tool in a separate window, GitHub Copilot's explicit-accept suggestion mode, and most standalone AI chat interfaces used alongside a composition surface are tier-1 architectures when operated as designed. The same products are tier-2 architectures when operated with inline completion, automatic acceptance, or aggressive copy-paste discipline.

The actionable directive: at tier 1, the deployment's actual tier is almost entirely determined by user behavior. Governance must focus on behavior rather than architecture. Tier-1 deployments in professional responsibility contexts should include explicit training and — where the stakes warrant — periodic audits of user practice. Tier-1 deployments in training or competitive IP contexts should include usage guidelines that make explicit what constitutes operating outside the tier.

Monday artifacts

CTO artifact: Architecture Decision Record for tier-1 deployment

Decision. We are deploying [system] at tier 1 because [specific business justification — professional responsibility, competitive IP, training objective, or explicit preference for human sole authorship]. We have explicitly rejected tier 2 because [specific reason joint authorship is undesired for this use case].

Preconditions verified. The tool is configured for suggestion-only operation (model output does not enter the composition surface by default). User training covers the tier-1 expectation — model output is reference material, not draft content to be adopted. Governance processes treat the sent artifact as human-authored, with no special provisions for AI-assisted content.

Preconditions NOT verified. Usually: user behavior in practice. Tier 1 cannot be architecturally enforced the way higher tiers can. Name this as an ongoing governance concern rather than pretending it is resolved.

What moves us up to tier 2. A determination that joint authorship is acceptable for this use case and that the benefits of integrated drafting outweigh the governance overhead.

What keeps us at tier 1. The specific professional responsibility, IP, or training rationale that justified the choice originally, verified as still operative on the quarterly review.

The tier-integrity watch. Tier 1's drift signal is behavioral, not metric-based. The indicator is usage pattern — if users are copying substantial portions of model output into their artifacts, the deployment is operating at tier 2 regardless of architecture. This is detectable through periodic sampling if the organization chooses to surface it, and not detectable otherwise. Name the sampling approach in the ADR if the deployment is in a context where tier-mismatch creates material exposure.

Owner. One human.

Review date. Quarterly minimum, with user-behavior audit proportionate to the rationale (frequent for professional responsibility contexts, lighter for training contexts).

Architect artifact: Configuration and usage-pattern monitoring

The architectural commitment at tier 1 is configuration, not instrumentation. The product must be configured such that model output does not enter the composition surface automatically — no inline completion, no automatic drafting, no one-click acceptance. The user's path to any model-generated text in the artifact must be deliberate.

Specific configuration points to verify, across most AI assistance products: inline completion disabled; automatic drafting features disabled; suggestion surface clearly separated from composition surface in the UI; any copy actions from suggestion to composition logged if the organization needs the audit trail.

Usage-pattern monitoring is optional but recommended in contexts where tier-mismatch creates material exposure. The monitoring is not the same as tier-2's diff logging — there is no sent-artifact diff to capture. What can be captured is suggestion-surface usage: how often users consult the suggestion panel, how long they spend there, how the usage correlates with the artifacts they subsequently produce. These signals do not prove operation at tier 2, but they can indicate deployments where behavioral review is warranted.

CPO artifact: User discipline and rationale communication

The tier-1-specific CPO concern is ensuring that the organizational rationale for tier 1 is communicated clearly enough that users understand why they are operating differently from the default.

In most AI-assistance contexts, tier 2 is the default user expectation. Users who have used similar tools elsewhere will expect inline drafting, automatic suggestions accepted with a keystroke, and the productivity gains that come with those workflows. Tier 1 deliberately removes those productivity gains in exchange for authorship clarity. Users who do not understand *why* they are operating at tier 1 will experience the constraint as arbitrary and will route around it — copying extensively from the suggestion surface, treating the tier-1 configuration as an obstacle rather than a choice.

The CPO's job is to make the rationale explicit and durable. In professional responsibility contexts, this means explicit reference to the professional framework that requires sole human authorship. In competitive IP contexts, this means clarity about what the organization is protecting and why. In training contexts, this means honesty that the organization has chosen slower authoring over faster authoring because the skill development matters. Users who understand the rationale will operate within the tier. Users who do not will operate at whatever tier maximizes their individual productivity, which is typically tier 2.

The internal-alignment concern from other tiers applies but in a simpler form. Sales, product, and support teams need to understand and be able to articulate why their organization has chosen tier 1 when tier 2 is the market default. This is not a vocabulary alignment problem (tier 2 is well-understood); it is a rationale alignment problem (why the organization is operating against the market default).

Forward references

Module 2 (evaluation) has limited relevance at tier 1 — there is no system behavior to evaluate because the system is not producing the artifact. Module 3 (incident response) covers tier-mismatch incidents: what to do when tier-1 governance has been assumed but users have been operating at higher tiers. Module 4 (procurement) covers contract language for tier-1 deployments, primarily around suggestion-data retention and the vendor's default configuration. Module 5 (organizational design) covers the training and communication responsibilities that carry the weight at this tier.

Tier 2: Draft-and-Edit

The recommendation, stated first

Tier 2 is the appropriate starting position for any use case where a human is going to author the final artifact anyway. That describes most current enterprise AI deployments — composition assistants, code copilots, content drafting — and the tier is correct for most of them.

The thing operators get wrong at tier 2 is not the architectural choice. It is underestimating the accountability shift that occurs the moment model-generated language appears in the sent artifact. Tier 2 feels like "the human is still doing the work." The legal and operational reality is that the human has become a *co-author* of content the model produced, with all the accountability consequences that co-authorship carries. Operators who deploy tier 2 with the mental model of tier 1 (suggestion-only) are exposed in ways they have not priced.

The architectural commitment

You are at tier 2 when the model produces output directly into the composition surface — the email body, the message draft, the code file, the document — and a human edits before sending or committing. The model's language enters the artifact. The human modifies it. The modified artifact is what the world sees.

The defining property is that *model-generated language enters the sent artifact, modified by the human*. This distinguishes tier 2 from tier 1 (where model output stays in a separate panel and the human composes from scratch) and from tier 3 (where the human approves the model's output binary-style without authoring it).

The shift from tier 1 to tier 2 sounds incremental. It is not. At tier 1, the human is the sole author of the sent artifact. At tier 2, the human and the model are joint authors. Every accountability property that depends on authorship — attribution, copyright, liability, professional responsibility — shifts when authorship becomes joint. Most deployments do not acknowledge this shift because it is invisible in the user experience.

What changes at tier 2

Attribution shifts from "human authored" to "jointly authored, human ratified." This is where the Air Canada question lives. The legal frame that is emerging treats the sent artifact as the organization's speech regardless of whether a human or a model drafted it. The human's editing does not remove the model's contribution; it supplements it. The organization is on the hook for both. This is the accountability shift most tier-2 operators have not internalized.

Auditability requires the diff. The distinctive logging requirement at tier 2 is not the model's output and not the human's final version — it is the *difference* between them. The diff is what reveals whether the human edited meaningfully or accepted substantially unchanged. Without the diff, you cannot distinguish a tier-2 deployment from a tier-3 deployment operating through an edit interface. Almost nobody logs this. The absence is not noticed until a dispute requires reconstructing what the model produced versus what the human sent.

Reversibility at tier 2 is mostly the normal reversibility of sent communications — but with a wrinkle. The wrinkle is that when the model's contribution to the artifact is revealed in adverse circumstances (a lawsuit, a regulator, a public embarrassment), the operator may need to distinguish what the model drafted from what the human authored, and that distinction is only possible if the diff was logged. Reversibility of the *artifact* is routine; reversibility of the *narrative* about who wrote it requires the diff.

Liability surface is genuinely evolving. The Air Canada ruling, the growing case law on AI-generated content in legal and medical contexts, the emerging professional-responsibility frameworks in multiple professions — all are tightening the position that organizations cannot disclaim model-generated language that appears in their artifacts. The specific frontier worth naming: *negligent misstatement* when model-generated content in an organization's artifact misleads a third party. The operator's defense cannot be "the model said it"; the defense must be "we exercised reasonable care in reviewing what the model said." Tier 2 is the tier where this defense operates, and it requires the editing to be genuine.

The regulatory angle: the Air Canada principle

Moffatt v. Air Canada is a 2024 British Columbia Civil Resolution Tribunal ruling. The facts: a passenger consulted Air Canada's website chatbot about bereavement fare policy; the chatbot provided incorrect information about the refund eligibility window; the passenger relied on the information; Air Canada subsequently denied the refund based on the actual (not stated) policy.

Air Canada's defense was notable for what it attempted: the airline argued that the chatbot was "a separate legal entity that is responsible for its own actions." The tribunal rejected this. The ruling's core finding — that Air Canada was responsible for information provided by its chatbot — is narrow on its face but expansive in implication. The airline could not disclaim the chatbot's statements merely because they were machine-generated.

The principle transfers beyond chatbots. The underlying doctrine is that organizations are responsible for the content they publish through their channels, regardless of whether the content was generated by a human employee, a contractor, or a model. This doctrine aligns with how courts have long treated other forms of organizational speech — a misleading advertisement produced by an agency is still the advertiser's misleading advertisement. The novelty is that the doctrine now explicitly applies to AI-generated content.

The implication for tier-2 deployments is direct. When a model produces content into an email, a document, a piece of code, or a customer-facing message, and a human sends it after editing, the organization is accountable for the full artifact — including the model's contribution. The human's editing does not sanitize the model's output; it incorporates it. Under the Air Canada principle, the question in an adverse event is not "did a human send it" but "was the organization's speech accurate and non-misleading." The human's role at tier 2 is to ensure the answer is yes; the editing is the mechanism for discharging that duty.

The U.S. surface is similar but less explicitly articulated. FTC guidance on AI-generated marketing content, emerging state-level disclosure requirements, and ongoing case law in defamation and negligent-misstatement contexts are all converging on the same position. Professional-responsibility frameworks in law, medicine, and accounting have moved faster — legal ethics opinions in multiple jurisdictions now explicitly treat AI-drafted content as the lawyer's work product for attribution and responsibility purposes.

The practical implication: at tier 2, the editing must be substantive. Accepting model output with cosmetic changes is operationally equivalent to signing off on whatever the model produced. Under the Air Canada principle, the organization has ratified the output by sending it. The distinction between tier 2 and tier 3 collapses operationally if the human's editing does not materially change the artifact.

When tier 2 is right

Three checkable conditions. All must hold.

The human has the competence to meaningfully evaluate and edit the model's output. The test: would the human have been capable of producing a correct artifact without the model's draft? If no, the editing cannot be genuine — the human lacks the substantive knowledge to know what to edit. Tier 2 deployments where the human is a generalist evaluating specialized content (a PM editing legal language, a junior editing specialist output) operate with a competence gap that the editing process cannot close. This does not mean tier 2 is wrong for such cases; it means the organization must be explicit that the output is being reviewed for surface characteristics, not for substantive correctness.

The editing is actually substantive, not cosmetic. The test: what fraction of model-generated language survives unchanged in the sent artifact? If the fraction is high — above what the organization considers "reviewed, not authored" — the deployment has drifted toward tier 3 and should be governed accordingly. The tier-integrity metric for tier 2 is the edit ratio, measurable if the diff is logged.

The organization accepts joint authorship of the artifact. The test: can the organization defend the sent artifact as its own speech if challenged? If the answer depends on arguing that the model authored the problematic portion, the organization has not accepted joint authorship — and under the Air Canada principle, that defense is unavailable. Tier 2 is only right if the organization is prepared to own the sent artifact in full.

If any of the three conditions fails, tier 2 is the wrong tier. Cases where competence is lacking should move to tier 1 (suggestion only, human composes from scratch). Cases where editing is not substantive should move to tier 3 (approve-and-send, with the appropriate governance discipline). Cases where joint authorship is not accepted should be reconsidered entirely.

The characteristic failure mode

The tier-2 characteristic failure mode is *silent drift toward tier 3 through editing atrophy*. Over time, users discover that accepting model output with minimal edits is faster than careful revision, produces artifacts that look acceptable to casual review, and does not trigger immediate consequences. The editing decays from substantive revision to cosmetic adjustment to rubber-stamp acceptance. The deployment migrates from tier 2 to a de facto tier 3 without any architectural change — and without any governance change, because the architecture still *supports* editing even when editing has stopped happening.

The Air Canada case is the canonical case study, worth treating structurally.

The chatbot at Air Canada was a tier-2 system in the broad sense — an AI-generated response displayed to users, with the organization in the position of having authored the response through the deployment. The tribunal did not distinguish between chatbot responses that had been reviewed by Air Canada staff and those that had not. From the tribunal's perspective, the distinction did not matter: the chatbot was the airline's speech regardless of review status. The ruling specifically rejected the "separate legal entity" defense.

The structural readings:

The editing step is legally invisible to the outside world. From the perspective of the recipient, the regulator, or the court, what matters is what the organization said. The organization's internal workflow for producing the statement — whether a human drafted it, a model drafted it and a human edited it, or a model produced it unedited — does not change the organization's responsibility for the statement. The editing at tier 2 is an internal quality control mechanism, not an external disclaimer.

"The model said it" is not a defense the legal system accepts. Air Canada's attempt to treat the chatbot as a separate entity failed for a reason that generalizes: organizational speech is organizational regardless of mechanism. Every tier-2 operator should internalize this. The editing exists to prevent the organization from saying something wrong; it does not exist to shift accountability if the organization does say something wrong.

The failure mode at Air Canada was upstream of the specific wrong statement. The chatbot was likely operating on a knowledge base that had not been rigorously maintained against the airline's actual policies. The wrong statement was a symptom of a system that was not structured to produce correct statements. Tier-2 operators whose systems have analogous structural weaknesses — models operating on outdated context, prompts that have not been audited, output that has not been spot-checked for accuracy — are producing Air Canada-shaped exposures continuously, whether or not any specific incident has yet surfaced.

The Air Canada ruling is a Canadian tribunal decision, not a Supreme Court ruling, and its binding authority outside British Columbia is limited. Its *persuasive* authority, however, is substantial — it articulates a principle that courts and regulators in other jurisdictions are converging on, and its reasoning has been cited in subsequent matters across multiple jurisdictions. The operator's question is not "does Air Canada bind my jurisdiction" but "is my jurisdiction likely to reach the same conclusion." For most developed jurisdictions in 2026, the answer is yes.

Vendor and architecture examples

Tier 2 is the modal architecture for currently-deployed enterprise AI. Microsoft Copilot in its default configuration across Office applications, GitHub Copilot, Google Workspace's Smart Compose and extended AI features, Notion AI, most composition assistants in sales and customer service platforms — these are tier-2 systems when the user accepts and edits model output in the artifact surface. The same products can be operated at tier 1 (suggestion panel, user composes from scratch) or tier 3 (approve-and-send, no meaningful editing) depending on the configuration and the user's behavior.

The actionable directive: at tier 2, the deployment's actual tier is determined by user behavior more than by product configuration. The same product can be tier 2 for a user who edits substantively, tier 3 for a user who rubber-stamps, and effectively tier 4 if downstream workflow automates sending without review. Governance at tier 2 cannot be purely architectural; it must include behavioral measurement and cultural reinforcement of the editing discipline.

Monday artifacts

CTO artifact: Architecture Decision Record for tier-2 deployment

Decision. We are deploying [system] at tier 2 because [specific business justification — typically: users are authoring artifacts anyway, and model assistance accelerates that authoring without replacing it]. We have explicitly rejected tier 1 because [specific reason inline drafting adds value beyond suggestion-only assistance]. We have explicitly rejected tier 3 because [specific reason users should be authoring rather than approving].

Preconditions verified. Diff logging is in place — the model's output and the user's final version are both captured, per artifact. Edit-ratio measurement is possible from the logs. Users have the competence to meaningfully evaluate and edit the model's output for their use cases. The organization has accepted that sent artifacts represent organizational speech, without distinction between human-authored and jointly-authored content.

Preconditions NOT verified. Usually: the competence-gap question. Most tier-2 deployments include some users evaluating content outside their primary expertise. Name this explicitly and describe the surface-review-not-substantive-review scope.

What moves us down to tier 1. A regulatory finding that tier 2's joint authorship creates unacceptable exposure for a specific use case; an incident revealing that competence gaps produced a material error.

What moves us up to tier 3. Sustained evidence that editing is not substantive in practice — that the deployment is functioning as tier 3 regardless of architecture — combined with a decision to govern it as tier 3 rather than continuing to claim tier 2.

The tier-integrity metric. Edit ratio — fraction of model-generated language that survives unchanged into the sent artifact. Measured per user, per artifact type, in aggregate. A high edit ratio (above whatever the organization considers "reviewed, not authored") signals drift toward tier 3 and triggers review.

Owner. One human. At tier 2, the owner carries responsibility for ensuring that joint authorship is acknowledged and governed as such.

Review date. Quarterly minimum; edit-ratio trends reviewed monthly.

Architect artifact: Diff logging and edit-ratio measurement

The diff is the load-bearing artifact at tier 2. Without it, the deployment is operating blind to its own tier.

Structurally, the diff log captures: the model's initial output, the user's final version at the moment of send/commit, the intermediate states if the user revised in multiple passes, and enough metadata to reconstruct the editing trajectory. The diff itself can be computed on demand; what must be stored is the anchor points (model output, final version, timestamps, user identity).

Edit-ratio measurement derives from the diff. The basic metric: *(tokens in final version that were not in model output) / (tokens in final version)*. More sophisticated metrics account for structural edits, semantic edits, and surface edits differently — but the basic metric is sufficient for tier-integrity monitoring. A falling edit ratio across a user population is the silent drift signal. A falling edit ratio within an individual user is a training or capacity signal about that user.

The diff log has a second function: it is the evidentiary artifact in any dispute about what the organization said versus what the model produced. The log should be retained for at least the statute of limitations relevant to the artifacts being produced, which for most commercial communications is several years. Most organizations retain this data for far shorter periods, often because they did not consider it evidence.

Privacy implications: the diff log contains user behavior data at fine granularity. Tier-2 deployments where users are employees operating in professional contexts generally have the legal basis for this collection. Tier-2 deployments in consumer contexts, where the user is the end-recipient of the artifact or a third party, raise collection questions that belong to privacy counsel. The architecture requirement is diff logging; the scope of logging is a governance question.

CPO artifact: Joint-authorship governance and editing-discipline programs

The tier-2-specific CPO concern is joint-authorship governance — ensuring that the organization has internalized the Air Canada principle and built operating discipline around it.

The first governance move is explicit organizational acceptance that sent artifacts represent organizational speech without distinction. This sounds abstract; it is operationally concrete. It means the organization does not have a mental category of "AI content" that is somehow less the organization's than "human content." It means policies for review, for retention, for liability allocation, for professional responsibility all treat AI-produced content as the organization's content. When procurement contracts with vendors allocate liability for AI-produced output, the allocation must be consistent with this position — vendor disclaimers that attempt to shift responsibility to the customer's users require pushback because they conflict with the organization's own accountability posture.

The second governance move is building editing discipline as an operating practice, not just an architectural feature. This is primarily a training and culture issue. Users need to understand why editing matters beyond their individual productivity — that the editing is the organization's discharge of its duty of care under the Air Canada principle, not a personal preference for authorship. The monthly edit-ratio review is the mechanism; the training program is what makes the mechanism meaningful. Organizations that deploy tier 2 without a training program are relying on users' individual judgment to carry accountability that belongs to the organization.

The third governance move is handling the competence-gap question honestly. Most tier-2 deployments include users evaluating content at the edges of their expertise. The organization's choice is: narrow the deployment to where users have substantive competence, supplement the deployment with expert review for the edges, or accept that portions of the deployment are effectively tier 3 and govern them accordingly. The wrong move — which is the common move — is to deploy widely, hope users will catch errors outside their expertise, and discover the gap only when an incident surfaces it.

The internal-alignment concern from earlier tiers applies here with a specific edge. Tier 2's invisibility to the outside world means the organization can underinvest in its governance without immediate consequence. The consequences surface later, in regulatory matters or litigation, when the organization's internal workflows are examined. The CPO's specific job is to ensure that the organization is not optimizing for short-term convenience at the cost of long-term accountability.

Forward references

The diff logging schema named above is the primary object Module 3 (incident response) treats at implementation depth, which absorbs the logging and reversibility infrastructure. Module 2 (evaluation) covers how to evaluate model-output quality when human editing is part of the feedback loop — a methodologically distinct problem from evaluating tier-3 or tier-4 systems. Module 3 also covers Air Canada-style incidents: what to do when model-generated content in a sent artifact surfaces as an organizational problem, including the editing-trajectory reconstruction from diff logs. Module 4 (procurement) covers contract language for tier-2 vendors, specifically the push-back required against vendor attempts to disclaim responsibility for model-generated content. Module 5 (organizational design) covers who owns editing-discipline programs — typically a shared concern between the product organization, learning and development, and the function whose content is being produced.

Tier 3: Approve-and-Send

The recommendation, stated first

Tier 3 is the default tier for new agentic deployments in 2026. Use it as the starting position and stay there until you have demonstrated, with production data, that reviewers are adding genuine judgment rather than rubber-stamping. Migrate down to tier 2 for high-stakes or highly-regulated work. Migrate up to tier 4 only after the readiness test in this module's tier-4 chapter has cleared.

The unusual thing about tier 3 is that the recommendation holds even though tier 3 is, in an important regulatory sense, the *most fragile* tier. The doctrine being contested most actively — whether pro-forma human approval satisfies "meaningful human involvement" under GDPR Article 22 and its emerging equivalents — sits directly on top of this tier. You are choosing a tier that regulators and courts are actively defining. You are choosing it anyway because the alternatives are worse for most use cases.

The architectural commitment

You are at tier 3 when the model produces a complete output, and a human has a binary gate — approve or reject — before the output is sent. The human does not typically edit the output; they approve it or decline it. The gate is per-action and synchronous: the action does not execute until the human acts.

The defining property is the *binary per-action gate*, synchronous with the action, without meaningful edit. This distinguishes tier 3 from tier 2 (where the human edits and authors the final artifact) and from tier 4 (where the human review is temporally separated from the action).

The architectural simplicity of tier 3 is deceptive. The tier looks like tier 2 with an extra convenience step. Operationally, it is closer to tier 4 than to tier 2 — the human has stopped authoring and started adjudicating, which is a cognitively different task with different failure modes, different auditability requirements, and different legal treatment.

What changes at tier 3

Three properties shift in ways that determine the tier's failure modes.

Attribution formally rests on the approver, but operationally dissolves at volume. In the legal frame, the human who clicked "approve" is accountable — they ratified the model's output. In the operational frame, when 500 approvals move through in an hour, the click is not meaningfully distinct from a rubber stamp. This is the tier where the legal frame and the operational frame diverge most sharply, and it is the gap regulators are actively closing. A reviewer who could not plausibly have considered the content they approved did not meaningfully approve it. The legal status of that approval is no longer settled.

Auditability requires a specific field most deployments don't capture: reviewer latency per action. At tier 3, the difference between meaningful review and rubber-stamping is visible in the logs only if you log *how long the reviewer spent before clicking approve*. Most systems log the approval event and the approver's identity. Few log the per-action review latency. Without that field, you cannot demonstrate — to yourself, to a regulator, to a court — that your tier-3 review was meaningful rather than pro-forma. The field is load-bearing for exactly the same reason that the policy-version identifier is load-bearing at tier 4: the eval harness and the regulatory defense both operate on it.

Reversibility is straightforward for the individual action and treacherous for the pattern. Any single approved action is reversible by the same mechanisms as a tier-2 action — retraction, correction, the normal apparatus of communication. The treacherous part is pattern reversibility. When a reviewer has been rubber-stamping for weeks, pulling the plug requires first *detecting* the pattern, and the pattern is invisible without the reviewer-latency field. You cannot rollback what you cannot see.

Liability surface expands in specific ways. GDPR Article 22 is the center of gravity, and it's worth going into depth.

GDPR Article 22 and the rubber-stamp problem

Article 22 prohibits decisions based solely on automated processing that produce legal or similarly significant effects, subject to exceptions. The historical assumption was that inserting any human in the decision loop satisfied the "not solely automated" requirement. Tier 3 architectures were the canonical response to this assumption — a human approves each automated recommendation, therefore the decision is not solely automated, therefore Article 22 does not prohibit it.

This assumption is being actively dismantled, and tier 3 is the tier the dismantling is happening to. Recent guidance and enforcement across EU data protection authorities — building on the *Schufa* ruling and its progeny — has begun to articulate what meaningful human involvement actually requires. The criteria that are emerging: the reviewer must have the *authority* to change the outcome, not just the formal power to click a button. The reviewer must have *access to the information* needed to evaluate the decision, not just the system's recommended action. The review *volume* must allow for genuine per-case consideration.

All three criteria cut against tier 3 as typically deployed. The first fails when approval queues are designed for throughput rather than deliberation; reviewers who face a thousand decisions per shift functionally lack the authority to meaningfully change any of them. The second fails when the review interface shows the recommended action without the underlying data, reasoning, or alternative outcomes the agent considered. The third fails the moment review volume exceeds what any human could genuinely consider — a threshold most enterprise tier-3 deployments cross quickly.

The implication for tier-3 architecture is specific. The deployment is defensible under Article 22 only if it is *designed* to satisfy the three criteria, not if a human happens to be in the approval seat. This means deliberately capping review volume per reviewer, deliberately surfacing the information needed for judgment, and deliberately empowering reviewers to escalate or reject without organizational friction. None of this is free. All of it is what makes tier 3 worth picking over tier 4 rather than a rubber-stamp layer on top of a tier-4 architecture in a trench coat.

The U.S. regulatory surface is moving in parallel, less formally. FTC guidance on AI-driven decisions in lending, housing, and employment increasingly scrutinizes whether human review is meaningful. State-level proposed legislation is converging on similar criteria. The December 2025 federal preemption EO complicates the state-level picture but does not resolve the federal one. Treat Article 22 as the bellwether doctrine; the U.S. surface will broadly follow, with sector-specific variations.

When tier 3 is right

Three checkable conditions. All must hold.

The review volume per reviewer allows genuine per-action consideration. This is the condition most often failed. The test: divide the expected action volume by the reviewer headcount by available work hours. If the resulting seconds-per-decision is below what a thoughtful reviewer would need to read the action, consult the underlying context, and form an independent judgment, tier 3 is mathematically incapable of delivering meaningful review in this deployment. You are at tier 4 with theater. Either cap volume, add reviewers, or move to tier 4 explicitly and treat the review accordingly.

The review interface surfaces what the reviewer needs to exercise judgment. The reviewer should see the recommended action, the underlying data it was based on, the agent's reasoning to the extent available, and — crucially — the ability to examine alternatives. "Approve this email?" is not enough information to approve an email. "Approve this email, shown alongside the prospect's history, the agent's reasoning, and the alternative framings it considered and rejected" is.

The escalation and rejection paths are frictionless. The reviewer must be able to reject without organizational penalty, escalate without bureaucratic delay, and surface patterns without reprisal. This is cultural as much as architectural. The architecture supports it (one-click reject, named escalation path) but the culture either ratifies it or hollows it. A tier-3 deployment where rejection is perceived as a performance problem for the reviewer is not meaningfully tier 3; it is tier 4 with extra steps.

If any of the three conditions fails, tier 3 is theater. Either fix the condition or acknowledge the tier you are actually at.

The characteristic failure mode

The tier-3 characteristic failure mode is *approval inflation* — the gradual erosion of the review step from meaningful adjudication to pro-forma click. It is not dramatic. It does not produce a single visible incident the way the 11x dialer did at tier 4. It produces a slow, invisible drift toward tier 4 behavior while retaining tier 3's architectural claim, until an external event — a regulator, a lawsuit, a journalist — forces examination of the logs and reveals that the approvals were never real.

The canonical case study is not a single AISDR incident. It is *Mata v. Avianca*, the 2023 SDNY sanctions ruling against the attorneys who submitted a brief containing fabricated case citations generated by ChatGPT. The case is worth reading in detail because the structural elements are precisely the tier-3 failure mode translated into the legal profession's version of an approval queue.

The attorneys had used ChatGPT to draft portions of a legal brief, including case citations. ChatGPT fabricated several of those citations — cases that did not exist. The attorneys filed the brief. The opposing counsel could not find the cited cases. The court asked the attorneys to produce copies. The attorneys returned to

ChatGPT, which produced fabricated text purporting to be the fabricated cases. The attorneys submitted this second round of fabrications to the court.

Judge Castel's sanction order made a specific finding that did structural work: the attorneys *continued to stand by the fake opinions* even after the issue was raised. The sanction was not for using AI-generated content; it was for ratifying it without verification. The approval gate — the attorneys' signature on the brief — was found to be insufficient because the review underlying the approval had not been performed.

The structural readings:

Tier 3's approval is legally insufficient if the review behind it is not performed. The court did not accept "I approved it" as a complete account. The court asked "what did you do between receiving the output and approving it," and the answer — nothing meaningful — was the basis for the sanction. Every tier-3 deployment is potentially subject to the same question. The logs need to support the answer.

The failure mode was not the model's hallucination — it was the human's ratification. The attorneys' fault was not in using ChatGPT. It was in treating the approval step as if it had been performed when it had not been. This is exactly the approval-inflation pattern at volume: individual approvals that have not meaningfully been performed, ratified as if they had, until external events force the question.

The sanction was individual, not organizational. The attorneys, not the firm, were sanctioned. This is worth internalizing for tier-3 deployments. The approver's name is in the logs. When the question is asked, it is asked of the named approver. Organizations that operate tier 3 without ensuring that their approvers understand this — that the approver's personal professional standing is what the approval puts on the line — are setting individuals up for liability the organization may or may not defend them against.

Mata is a legal-profession case but the pattern transfers. Any tier-3 deployment where approvers are processing high volumes of agent output under organizational pressure to keep throughput high is reproducing the Mata conditions in a different industry.

Vendor and architecture examples

Tier 3 is the architectural position the market has settled on as the safe middle ground for early 2026 agentic deployments. The major sales engagement platforms (Outreach, Salesloft, HubSpot) offer tier-3 approval queues as a configurable mode. Most compliance-adjacent agentic products (Kasada, Harvey for legal, certain healthcare workflow agents) are deployed at tier 3 by default. Most enterprise customer service deployments that claim "human-in-the-loop" operate at tier 3, though the "loop" is often tighter in the marketing than in the architecture.

The actionable directive: when evaluating a tier-3 product, do not ask the vendor whether the product has an approval gate. Every vendor will say yes. Ask three specific questions. What is the expected review volume per reviewer per hour in a typical deployment? Does the approval interface surface the underlying data and the agent's reasoning, or only the recommended action? Can the reviewer latency per action be logged and audited? If the answers are unsatisfactory on any of the three, the product is architected for tier 4 with a tier-3 veneer, regardless of the contract language.

Monday artifacts

CTO artifact: Architecture Decision Record for tier-3 deployment

Decision. We are deploying [system] at tier 3 because [specific business justification]. We have explicitly rejected tier 2 because [specific reason full human authorship is not required or not feasible]. We have explicitly rejected tier 4 because [specific reason synchronous per-action review is warranted — regulatory exposure, blast-radius concentration, reviewer-capacity availability].

Preconditions verified. Review volume per reviewer is below the threshold at which meaningful per-action consideration becomes mathematically impossible; the specific threshold for our use case is [N seconds per decision minimum]. Review interface surfaces the recommended action, underlying data, and agent reasoning. Reviewer-latency-per-action is logged. Escalation and rejection paths are frictionless and culturally ratified.

Preconditions NOT verified, and why we are proceeding anyway. List them. If this list is empty, the list is wrong.

What moves us down to tier 2. A specific regulatory finding, a specific incident type, a specific shift in risk profile that warrants full human authorship.

What moves us up to tier 4. A demonstrated track record of reviewer judgment catching meaningful errors, combined with reviewer-capacity data showing that synchronous review is creating throughput constraints that the business cannot absorb.

The approval-inflation watch. A named metric and a named owner. *Reviewer latency per action, tracked weekly, with a specific lower bound below which the deployment is automatically flagged for review.* This is the tier-3 equivalent of the dialer check.

Owner. One human.

Review date. Quarterly minimum.

Architect artifact: Logging schema and review-interface spec

The tier-3 logging schema adds three fields to the tier-4 baseline: reviewer identity per action, reviewer latency per action (time between action presented to approver and approval click), and review-interface state at time of approval (what the approver could see when they clicked).

Reviewer latency is the load-bearing field. Without it, approval inflation is undetectable. With it, approval inflation is a simple dashboard query: distribution of reviewer latency over time, per reviewer and in aggregate, with automated alerting when distributions compress toward zero. When the distribution collapses, the deployment has drifted to tier 4 regardless of what the architecture claims.

The review-interface spec: the approver's screen at decision time must include the agent's recommended action, the underlying data the recommendation was based on, the agent's reasoning to the extent exposed by the model, and explicit alternatives the agent considered and rejected. Interfaces that show only the recommended action are architected for throughput, not judgment. The approver cannot exercise judgment about what they cannot see.

The rejection primitive: a one-click reject that does not require justification text. Requiring justification text for rejection creates asymmetric friction — approve is easy, reject is costly — which produces approval inflation by design. Justification should be requested for patterns (high reject rates, specific reject reasons trending up) but not for individual rejections.

CPO artifact: Reviewer protection and approval-quality governance

The tier-3-specific CPO concern is reviewer protection and approval quality. These are two sides of the same problem.

Reviewer protection means the organization must make rejection and escalation professionally safe. If rejection is perceived as a performance problem for the reviewer, tier 3 collapses into tier 4 within months. The CPO's job is to ensure that the performance systems, the product roadmap, and the internal communications treat high reject rates as potentially a feature (the reviewer is catching things) rather than a defect (the reviewer is slow). This is a governance move, not a technical move.

Approval quality governance means building the feedback loop from downstream outcomes back to review patterns. When an approved action produces a bad downstream outcome, the post-mortem should examine not only the action but the review. Which reviewer approved it? What was their latency? What did their interface show them? This is not about blaming reviewers; it is about closing the loop so the review architecture improves over time. Most organizations never build this loop. Without it, the reviewer-latency field is just a number in a log; with it, it becomes the lever that keeps tier 3 operational.

The CPO should also own the Mata-style individual-vs-organizational liability question internally. Approvers should know what they are putting their name on. Organizations should know whether and how they will defend named approvers in adverse events. These are governance questions, not legal questions, and they belong to the product executive.

Forward references

The reviewer-latency field named above is the object Module 2's eval harness uses for tier-3 systems — the harness evaluates reviewer judgment quality against downstream outcomes, and it operates on the latency field plus the interface-state field together. Module 3 covers the review-interface spec at implementation depth, including the logging and reversibility infrastructure. Module 3 also covers incident response for tier-3 systems, specifically the approval-inflation detection pattern. Module 4 covers contract language — including vendor representations about expected review volumes and the discrepancy between the product's designed review capacity and the customer's actual reviewer headcount, which is a common and contentious procurement question. Module 5 covers reviewer org design, which sits between the product organization and operations and is often orphaned.

Tier 4: Supervised Execution

The recommendation, stated first

For new agentic deployments in 2026, start at tier 3 (approve-and-send) and migrate to tier 4 only after you have demonstrated, with production data, that reviewers are making meaningful distinctions rather than rubber-stamping, and that your reversibility architecture works under real failure conditions.

Vendors will push you toward tier 4 for the demo effect. Resist. The migration from 3 to 4 is straightforward when you're ready; the migration backwards from 4 to 3 after a public incident is expensive, visible, and damages the product roadmap for quarters.

The rest of this chapter is the reasoning.

The architectural commitment

You are at tier 4 when your system takes externally-visible actions without a per-action human gate, and a human reviews what happened afterward, typically through dashboards, sampled audits, or exception queues.

The defining property is not autonomy. The defining property is that **the human review is temporally separated from the action.** At tier 3, the human reviews before send. At tier 4, the human reviews after send, if at all, and typically on a sample.

This sounds like a minor shift. It is not. Moving the human review from *before* to *after* changes every accountability property qualitatively. Most operational failures attributed to "AI agents" are tier-4 failures specifically — and most organizations deploying at tier 4 believe they are deploying at tier 3 or tier 5.

What changes at tier 4

Two properties change in ways that drive everything else. I'll treat these at length. Attribution and liability surface also change, but in ways that are mostly shared with tier 5 — I'll flag the distinctions briefly and carry the detailed treatment forward to Module 6 (regulatory posture).

Auditability balloons and almost nobody meets the new requirement. At tiers 1–3, the human's approval or edit is itself the primary audit record. At tier 4, there is no human record — the only record of why the system did something is the system's own decision trace. That trace has to contain inputs, retrieved context, tool calls, intermediate reasoning, the final output, and the downstream effects. Shakudo's 2026 report on production agent failures found that the majority of enterprise deployments lack structured evaluation harnesses, distributed tracing, or systematic failure classification — which means when something breaks, teams cannot determine whether the failure originated in the prompt, the model, a tool integration, or the orchestration layer. That's not a tooling gap. It's an accountability failure that becomes visible only when a specific action is challenged and the logs cannot reconstruct why it happened.

Reversibility becomes engineering, not policy. Undoing an email sent at 2am requires a retraction protocol — and retractions damage sender reputation, confuse recipients, and in regulated industries are compliance events in themselves. Undoing a database write requires transactional design that includes the agent's actions as first-class citizens. Undoing a third-party API call requires compensating actions that may or may not exist. At this tier, reversibility is not a property of the system you bought; it is a property of the architecture you built around the system you bought. Most organizations at tier 4 have not built that architecture, which means their reversibility is worse than they think, and they discover this during an incident rather than before one.

Attribution changes too — the deploying company owns the outcome without a named approver per action, which means "who authorized this" has no natural answer and must be constructed from the logs. If the logs don't support the construction, the attribution gap is a liability. Liability surface expands to include FTC enforcement posture, sector-specific rules, and the continuing evolution of GDPR Article 22's treatment of "decisions based solely on automated processing," where regulators and courts are increasingly scrutinizing whether sampled post-hoc review qualifies as meaningful human involvement. The specific Article 22 treatment is worth going deeper on.

GDPR Article 22 and the meaningful-human-involvement question

Article 22 prohibits decisions based "solely on automated processing" that produce legal or similarly significant effects on individuals, subject to specific exceptions. For years, operators assumed that inserting any human review satisfied the "not solely automated" requirement. That assumption is being actively eroded.

Recent regulatory guidance across EU data protection authorities has begun to distinguish between rubber-stamp review and meaningful human involvement. The criteria regulators are signaling include: whether the reviewer has the authority to change the outcome, whether the reviewer has access to the information needed to evaluate the decision, and whether the review volume allows for genuine consideration per case. A reviewer approving hundreds of decisions per hour cannot, by construction, satisfy the third criterion.

Tier 4 sits precisely in the regulatory blast zone of this doctrine. The sampled post-hoc review characteristic of tier-4 deployments is the pattern Article 22 interpretation is tightening around. For any AISDR use case that touches decisions about individuals — and most do, to the extent that targeting and messaging personalization are themselves decisions about individuals — the operator should assume that Article 22 applies, that post-hoc sampled review will not satisfy the meaningful-involvement test, and that the defensible architectures are either tier 3 (per-action approval) or tier 5 (bounded autonomy with documented constraints that functionally narrow the decision space).

The rest of the tier-4 liability surface — FTC, sector regimes, state data-broker rules, EU AI Act classification — is treated in Module 6 (regulatory posture). The contract language your MSA should carry to push vendor disclaimers against these regulatory surfaces is treated in Module 4 (procurement).

When tier 4 is right

Three checkable conditions. All must hold.

Actions are high-volume and low-individual-consequence, such that no single action produces a catastrophic outcome and the review cadence can catch systemic problems before aggregate blast radius grows large. The test is not "we do many actions" — it's "one wrong action does not produce a headline."

Outputs are verifiable after the fact by the reviewer without reconstructing the system's reasoning from scratch. If verification requires the reviewer to redo the work the agent did, the review is not catching what tier 4 needs it to catch.

Reversibility is designed in, not assumed. Retraction protocols for communications, transactional boundaries for data writes, compensating actions for every third-party integration, rollback paths for both individual actions and policy changes. If these are not built, the organization is at tier 4 in name and somewhere worse in practice.

If any of the three conditions fails, tier 4 is the wrong tier regardless of what the vendor's marketing calls the product.

The characteristic failure mode

The tier-4 characteristic failure mode is not "the system did the wrong thing." It is that *the system did the wrong thing and nobody noticed in time to reverse it*, because the review architecture was aspirational rather than operational.

The 11x Inc. case is the fullest public instance of this failure mode in the AISDR domain. The structural elements — not the full narrative; the narrative is in TechCrunch's March 2025 reporting and elsewhere:

11x marketed an AI SDR ("Alice") as digital workers replacing multiple full-time employees. ZoomInfo ran a one-month trial in early 2025, found the product performed significantly worse than its own human SDRs, and churned. 11x continued claiming ZoomInfo as a customer on the website, in sales calls, and on the AI phone dialer for months. The website logo was removed around March

6, 2025, after press inquiry. *The AI phone dialer continued asserting ZoomInfo as a customer after the website was updated.* Airtable followed a parallel pattern. Customers reported hallucinations, product failures, and an inability to exercise contractual break clauses.

Three structural readings the reader should extract.

Tier-mismatch produces the dispute. The product operated at tier 4 — autonomous send, post-hoc review by the customer. It was marketed at tier 5 or 6 — self-directing digital workers requiring minimal supervision. Customers bought the marketed tier and discovered the operational tier, which required more oversight than they had allocated and produced fewer qualified meetings than the marketing implied. The TechCrunch story is the visible tip of this gap. Every agentic AI product carries some version of this risk; the question is how wide the gap is and how the product handles closure.

The dialer detail is the canonical tier-4 reversibility failure. When 11x's website was updated to remove the ZoomInfo logo, the AI phone dialer continued asserting ZoomInfo as a customer. Someone changed the authoritative record. The deployed autonomous system did not reflect the change. This is not a bug in the ordinary sense — it is the signature failure mode of tier-4 systems that were built without treating the system's own assertions as reversible state. Every organization deploying at tier 4 should read the dialer detail and ask: *when we update a fact in our authoritative systems, how long until every autonomous surface reflects the update?* If the answer is "I don't know," the organization is not at tier 4. It is somewhere worse, and doesn't know it.

The incorrect lesson the market took. Multiple incumbent vendors used the 11x case to argue that AI SDRs don't work as a category. That overreads the evidence. The case establishes that deploying at tier 4 while marketing at tier 5 produces catastrophic customer outcomes. It does not establish that tier 4 is wrong for the category.

The vendor landscape is not the useful takeaway. The useful takeaway is the principle: marketed tier and operational tier must match, and when they don't, customer trust collapses faster than any other business metric.

Vendor and architecture examples

Operational-tier-4 deployments are the modal AISDR architecture as of early 2026. The specific tier any given product occupies depends on the configuration the customer chose; most mainstream AISDR products can be operated at lower tiers with explicit setup, and most are operated at or near tier 4 by default.

The actionable directive: when you evaluate an AISDR vendor in 2026, assume operational tier 4 regardless of marketed tier. Run the diagnostic in Part 5 on the actual deployment, not on the sales demo. If the operational tier you measure differs from the tier the vendor claimed in the contract, that is a material discrepancy and belongs in the procurement conversation Module 4 covers in full.

Monday artifacts

CTO artifact: Architecture Decision Record for tier-4 deployment

One page. If it's longer, it isn't load-bearing.

Decision. We are deploying [system] at tier 4 because [specific business justification]. We have explicitly rejected tier 3 because [specific reason review cadence cannot be synchronous]. We have explicitly rejected tier 5 because [specific reason we are not yet ready to declare and enforce operational constraints].

Preconditions verified. Distributed tracing in place. Full decision traces logged per action. Retraction protocols built. Transactional boundaries defined for data writes. Compensating actions defined for every third-party integration. Review cadence matched to blast-radius-per-unit-time. Regulatory surface mapped, including Article 22 exposure.

Preconditions NOT verified, and why we are proceeding anyway. List them. Every tier-4 deployment has unverified preconditions; pretending otherwise produces the 11x case. If this list is empty, the list is wrong.

What moves us down to tier 3. A specific incident type, a specific regulatory change, a specific reviewer-capacity threshold. Named in advance.

What moves us up to tier 5. A specific demonstrated track record with reviewers catching meaningful errors, a specific boundary architecture ready for declaration.

Owner. One human who carries the decision.

Review date. Quarterly minimum.

Architect artifact: Logging schema and rollback spec

The load-bearing fields in the logging schema: action ID (globally unique, referenced in downstream systems), full inputs record, reasoning trace to the fullest extent the model exposes, the action itself, downstream effects linked by action ID, *policy version identifier*, and sub-second timestamp.

The policy version identifier is the field the eval harness in Module 2 operates on — it's what allows you to evaluate behavior per policy version rather than per action, which is what makes drift detection tractable. Without it, you cannot tell whether a bad action was a bad sample from a good policy or a signal of policy drift. This is why it's load-bearing despite looking incidental.

The rollback spec defines four things per action category: the retraction primitive (what specifically undoes this action — if none exists, the category is not tier-4-eligible), the detection-to-rollback latency target (the SLA that matters most in an incident and that most teams have never measured), the policy-rollback procedure (how you revert a bad policy update and how you verify it reverted), and the dialer check (for every authoritative record the agent reads from, document maximum propagation latency between the record changing and all agent surfaces reflecting the change; if the answer is "unknown," that's the first thing to fix).

CPO artifact: Brand-risk and retraction-latency playbook

The tier-4-specific CPO concern is not disclosure. It is brand risk from autonomous action the CPO didn't authorize and can't retract fast enough.

Build the retraction-latency playbook before you need it. Name the brand-sensitivity tiers: which customer segments, which channels, which message types carry outsized reputational risk if a wrong action reaches them. For each tier, set the maximum acceptable detection-to-retraction latency. These numbers are the operational translation of brand risk, and they are the numbers the product org should own.

Align the marketed tier, the operational tier, and the internally-understood tier. The 11x case is, among other things, a case of internal mental-model divergence — sales, product, and support operating with different mental models of the system's autonomy until external reporting forced resolution. The CPO's specific job at tier 4 is to ensure these three mental models are the same, and that the escalation path when they diverge runs through the product org rather than through PR after the fact.

Forward references

The policy version identifier in the tier-4 logging schema is the primary object Module 2's eval harness operates on — behavior is evaluated per policy version, which is why the identifier is load-bearing despite looking incidental. Module 3 covers the full logging schema at implementation depth, including the downstream-effects graph and distributed trace propagation. Module 3 also covers incident response, specifically the retraction-latency and policy-rollback procedures named above. Module 4 covers contract language — including vendor disclaimers, Article 22 allocation, and the "operational tier vs. marketed tier" discrepancy as a material term. Module 5 covers who owns the tier-4 ADR in the enterprise org chart; short version — not a committee.

Tier 5: Bounded Autonomy

The recommendation, stated first

Tier 5 is the target architecture for mature agentic deployments — but it is not a starting position. Migrate to tier 5 from tier 4 only after you have built, deployed, and operationally verified the two things tier 5 rests on: a declared constraint architecture that narrows the decision space in defensible ways, and an exception-handling regime that treats boundary events as first-class operations rather than edge cases.

The common failure at this tier is not operating at tier 5 poorly. The common failure is *claiming* to operate at tier 5 while actually operating at tier 6 — running an agent without meaningful constraints, pointing at aspirational boundary language in a design doc, and calling the combination "bounded autonomy." This is the tier where the gap between claimed architecture and actual architecture is widest, and where the gap is hardest to detect from the outside because the architectural intent sounds sophisticated.

The architectural commitment

You are at tier 5 when the model acts freely within *declared, operationally-enforced constraints* and requests human involvement only when it hits a boundary. The declared constraints are documented, version-controlled, and machine-enforced — not described in a design document. The exception-handling path is built, staffed, and measured.

The defining property is the *declared constraint architecture*. At tier 4, the system acts autonomously and a human reviews afterward. At tier 5, the system acts autonomously within boundaries it cannot cross, and a human handles the boundary events. The shift is that *oversight moves from the actions to the boundaries.* The human is no longer reviewing what the agent did; the human is adjudicating where the agent is allowed to operate.

This is why tier 5 is structurally different from tier 4 rather than just "tier 4 with guardrails." The accountability object has changed. At tier 4, the accountability object is the action. At tier 5, the accountability object is the boundary.

What changes at tier 5

Attribution now includes the boundary designer, the boundary approver, and the boundary operator — three distinct roles that are often collapsed. The boundary designer decides what constraints the agent operates within. The boundary approver ratifies those constraints at the organizational level, typically at the level where risk appetite is set. The boundary operator maintains the constraints in production, updating them as conditions change. At tier 4, attribution is muddy because no human authorized the specific action. At tier 5, attribution is *clearer per action* (the action was authorized by being inside the boundary) but newly complex around the boundary itself. When something goes wrong, the question is no longer "who approved this action" but "who designed the boundary that permitted it, who approved the boundary, and who maintained it."

Auditability adds a dimension: boundary evaluations, not just actions. Every time the agent considers an action and decides it is inside the boundary, that evaluation is itself an accountability event. The logs must capture not only what the agent did but what the agent considered doing, what it decided not to do, and why. This sounds excessive until the first incident. The first incident is almost always a boundary-evaluation failure — the agent took an action that was technically inside the declared boundary but that the boundary designer had not intended to permit. Reconstructing that failure requires the boundary-evaluation log, not the action log. Almost nobody captures this at deployment.

Reversibility splits into two distinct operations: action rollback and boundary rollback. Action rollback inherits tier 4's apparatus — retraction, compensating actions, transactional boundaries. Boundary rollback is a new operation. When a boundary update on Tuesday contributed to a bad outcome on Wednesday, rolling back to Monday's boundary is a specific, versioned operation with specific propagation semantics. The boundary is code. It needs the same rigor as production code: version control, change review, canary deployment, rollback procedures. Most organizations deploying at tier 5 treat boundaries as configuration, which is the wrong mental model. Boundaries are policy, and policy is code.

Liability surface shifts toward duty-of-care analysis. At tiers 2 through 4, liability analysis focuses on the action: was the output permissible, was the review meaningful, was the blast radius contained. At tier 5, liability analysis focuses on the *boundary design*: was the constraint architecture reasonable, did it reflect industry standards, was it updated when circumstances warranted. Plaintiffs' counsel will argue the boundary was negligently designed. Defendants will argue the boundary reflected reasonable care. This is unfamiliar terrain for most operators — the analysis is closer to product liability than to traditional tort — and the case law is still forming.

The regulatory angle: boundaries as satisfaction of meaningful human involvement

The Article 22 doctrine that cuts so sharply against tier 3 and tier 4 relates to tier 5 differently. The meaningful-human-involvement requirement can potentially be satisfied at tier 5 *through the boundary architecture itself*, provided the architecture is designed and maintained to that standard.

The argument: if the boundary is designed by humans, reviewed by humans, approved at a level commensurate with the risk, updated by humans as conditions change, and enforced such that the agent cannot act outside it, then meaningful human involvement has been inserted *into the decision space* rather than *into each decision*. The human is not reviewing each action; the human has bounded the range of permissible actions in a way that reflects deliberative judgment.

This argument is not yet settled law. Regulators have not explicitly endorsed it. Plaintiffs' counsel will resist it. But it is the strongest defensible position for automated decision-making at scale, and it is the direction mature regulatory thinking is moving. The practical implication is that tier 5 is the tier where GDPR Article 22 compliance becomes architecturally achievable for high-volume applications — provided the boundary architecture is genuinely what the defense claims it is, rather than aspirational language in a slide deck.

This is also why the failure mode of "claiming tier 5 while operating at tier 6" is regulatorily dangerous. An operator who relies on tier 5's Article 22 position without actually having built the architecture is not merely architecturally weak — they have asserted a regulatory defense they cannot support under examination.

When tier 5 is right

Three checkable conditions. All must hold.

The constraint architecture is declared, version-controlled, and operationally enforced — not described. The test: can you produce the current boundary specification, the history of changes to it, the approver of each change, and the mechanism that prevents the agent from acting outside it? If any of the four is missing, the boundary does not operationally exist. It is aspirational.

Exception handling is built, staffed, and measured. The test: when the agent hits a boundary, what happens? Who handles the exception? What is the median and 95th-percentile handling time? What fraction of exceptions are resolved within the defined path versus escalated versus abandoned? If you cannot answer these questions from your monitoring, exception handling is a story you tell, not a process you run.

The boundary-design process is proportionate to the risk. The test: was the boundary designed by people with the authority and information to define the organization's risk appetite? At scale and at material risk levels, boundary design is an executive function, not a developer function. Boundaries designed by engineers in a sprint are unlikely to reflect the organization's actual risk posture. Boundaries approved by risk committees but never updated are unlikely to reflect current conditions. Proportionality means the design process matches the stakes.

If any of the three conditions fails, you are operating at tier 6 and calling it tier 5. Either meet the conditions or acknowledge the tier you are actually at.

The characteristic failure mode

Tier 5 fails in two characteristically different ways, and they deserve separate treatment because the remedies are opposite.

Failure mode one: the boundary is too loose. The agent acts within the declared boundary and causes harm that the boundary was intended to prevent. This is a *design* failure — the boundary did not reflect the organization's actual risk appetite, either because it was set too permissively at deployment or because it was not tightened as conditions changed. The remedy is tightening the boundary and improving the design process.

Failure mode two: the boundary is too tight. The agent hits the boundary constantly, every interaction becomes an exception, the exception queue overflows, reviewers stop handling exceptions meaningfully, and the system degrades to tier 3 or tier 4 under the pressure. This is a *calibration* failure — the boundary was conservative to the point of preventing the agent from doing useful work. The remedy is loosening the boundary and improving the exception-handling capacity.

The failure modes are inverse, but they share a common root: the boundary was not calibrated to actual operating conditions. The organization that defaults to tight boundaries thinks it is being safe; it is actually pushing the accountability problem into the exception queue. The organization that defaults to loose boundaries thinks it is being pragmatic; it is actually operating at tier 6 while claiming otherwise.

The canonical case study at tier 5 is not a catastrophic single-incident failure. It is a successful migration — the Salesforce Agentforce trajectory from 2024 through 2026.

Salesforce launched Agentforce in 2024 with strong tier-6 positioning: autonomous agents, self-directing workflows, systems that could reason, decide, and act without human intervention. The 2025 deployment experience did not match the positioning. By late 2025, Salesforce publicly shifted the narrative. The language moved from probabilistic execution toward *guardrails, deterministic automation, and data-first design*. This was not a model improvement. It was an architectural correction — a retreat from tier 6 toward tier 5, with tier 5's defining feature (declared, enforced constraints) taking the positioning weight.

The structural readings:

Mature vendors migrate down the gradient, not up. The industry narrative in 2024 framed the trajectory as agents becoming more autonomous over time. The actual 2025 trajectory was agents becoming more constrained. The vendors that survived the year did so by adding constraints, not by removing them. This is an important corrective to the implicit assumption that tier 6 is the mature state and lower tiers are training wheels. The evidence suggests tier 5 is the mature state and tier 6 is an immature marketing position that the market is correcting.

The correction was narrative as well as technical. Salesforce's shift was visible in public-facing language before it was visible in product changes. This is worth internalizing. When a major vendor changes how they talk about autonomy, that change usually leads product architecture by one to two quarters. Watch the language, not the roadmap.

"Guardrails" is tier 5's term of art, and it is doing heavy lifting. The word papers over real architectural distinctions. Some guardrails are boundary constraints (tier 5). Some are pre-send filters (effectively tier 3 in disguise). Some are post-hoc audits (tier 4). When a vendor says "guardrails," the operator's job is to ask which kind, specifically, and how they are enforced. "Guardrails" without architectural specificity is marketing, not design.

The cross-domain supplement worth brief note, flagged as such: Cruise's 2023 robotaxi permit suspension is the clearest public case of tier-5 boundary failure in a non-AISDR domain. The Cruise system operated within declared operational boundaries. A pedestrian was struck. The system then dragged the pedestrian twenty feet before stopping — a boundary-evaluation failure where the agent's model of "stop safely" did not include the correct handling of pedestrians trapped underneath. The California DMV suspension cited Cruise's delayed and incomplete disclosure — which is the tier-5 exception-handling failure. The boundary was too loose in a specific edge case; the exception handling failed when the edge case occurred. Both failure modes in one incident. Brief reference, not spine; the AISDR translation is straightforward.

Vendor and architecture examples

Tier 5 is claimed by most enterprise agentic products in 2026 and operated by significantly fewer. The gap between claim and operation is the category's most pervasive procurement risk.

Architecturally credible tier-5 deployments typically share three visible features: explicit constraint specifications in documentation (not marketing), published exception-handling SLAs with observed performance data, and versioned boundary change history available to customers. Products that offer the first without the second and third are claiming tier 5 while operating at tier 6.

The actionable directive: when evaluating a vendor claiming tier 5, ask for the boundary specification, the exception-handling metrics, and the change history. If the vendor cannot produce the first, they are at tier 6. If they produce the first but not the second and third, they have the architectural intent without the operational discipline — which is the most common and most dangerous position because it looks like tier 5 from the slide deck.

Salesforce Agentforce's 2026 positioning, Anthropic's own deployed agentic products, and a handful of vertical enterprise agent vendors (specifically those in regulated industries where the regulatory pressure has forced genuine tier-5 discipline) are the clearest operators at the time of writing. The list is short and will change; trust the architectural criteria, not the vendor list.

Monday artifacts

CTO artifact: Architecture Decision Record for tier-5 deployment

Decision. We are deploying [system] at tier 5 because [specific business justification — typically: synchronous review is infeasible at volume, and we have both the maturity to design constraints and the engineering capacity to enforce them]. We have explicitly rejected tier 4 because [specific reason the boundary architecture adds value beyond post-hoc review]. We have explicitly rejected tier 6 because [specific reason unbounded autonomy is not an acceptable architectural position for this use case].

The boundary specification. Not described here. *Referenced here.* The boundary specification is a separate, version-controlled artifact with its own approval chain. The ADR names the boundary spec, its version, its owner, and its approval date. If the boundary spec does not exist as a separate versioned artifact, the deployment is not at tier 5.

Preconditions verified. Boundary specification exists, is version-controlled, and is operationally enforced (not documented only). Exception-handling path is built, staffed, and has published SLAs. Boundary-change procedure is defined, with named approvers at appropriate authority level. Boundary-rollback procedure is defined and tested. Boundary-evaluation logging is in place.

Preconditions NOT verified, and why we are proceeding anyway. List them. At tier 5, the most commonly unverified precondition is proportionate boundary design — boundaries were set by developers without risk-committee review. Name this explicitly if it applies.

What moves us down to tier 4. A specific incident demonstrating that the boundary architecture is not preventing outcomes it was designed to prevent, a specific regulatory development, or a demonstration that exception-handling capacity cannot scale with the system.

What moves us up to tier 6. There is no legitimate migration from tier 5 to tier 6. If the pressure is to remove constraints rather than add them, the organization is misreading the maturity trajectory.

The tier-integrity metric. Boundary-exception handling time — median and 95th-percentile time between an exception triggering and meaningful handling. This metric is the tier-5 equivalent of the dialer check and the reviewer-latency field. When it degrades, the deployment is silently becoming either tier 4 (exceptions stop being handled meaningfully) or tier 6 (the boundary is being loosened informally to reduce exception volume).

Owner. One human. At tier 5, the owner carries responsibility for the boundary design, not just the deployment decision.

Review date. Quarterly minimum; boundary specification reviewed on a schedule proportionate to risk level.

Architect artifact: Boundary specification and exception-handling architecture

The boundary specification is the load-bearing artifact. Its format matters less than its properties: it must be machine-readable by the enforcement layer, human-readable by the approvers, version-controlled, diff-able across versions, and comprehensively enforced.

Structurally, a boundary specification defines: permitted action types (whitelist, not blacklist — a whitelist fails closed when an unexpected situation arises; a blacklist fails open); permitted targets (what systems, accounts, data scopes the agent can operate against); permitted magnitudes (spend limits, volume limits, rate limits, blast-radius limits); permitted contexts (temporal constraints, approval prerequisites for specific action types); escalation triggers (what conditions move an action from autonomous to exception).

The enforcement layer sits between the agent and its action space. Every action the agent attempts passes through the enforcement layer; actions outside the boundary are blocked, not permitted-with-warning. Soft enforcement (log-only) is tier 6 in disguise.

The exception-handling architecture defines: the exception queue (where exceptions go when raised), the handler roster (who is on call to process exceptions), the SLA (handling time by exception severity), the disposition options (approve, reject, modify, escalate), and the feedback loop (how exception patterns feed back into boundary refinement). The loop is the part most often omitted. Without it, the boundary stays static while conditions drift, and the exception queue reveals the drift only at the point of failure.

Boundary change management follows the same discipline as production code: version control, change review, canary deployment where feasible, rollback procedures, post-change monitoring. The operational equivalent of "deploying a new boundary" should feel like deploying a new service, not editing a config file.

CPO artifact: Boundary governance and risk appetite alignment

The tier-5-specific CPO concern is boundary governance — ensuring that the constraint architecture reflects the organization's actual risk appetite, that the approval chain operates at commensurate authority, and that the boundary evolves as conditions change.

The risk-appetite alignment question is the most underserved in current practice. Most tier-5 boundaries are set by the engineering team that built the agent, often under schedule pressure, often without input from the risk function, legal, compliance, or the executive team responsible for the product's market positioning. This produces boundaries that reflect engineering feasibility, not organizational risk appetite. The consequence is that the first incident surfaces a mismatch: the agent operated within its declared boundary, the outcome was unacceptable to the organization, and the root cause is that the boundary was never approved at the level where the organization actually defines what is acceptable.

The CPO's specific job at tier 5 is to ensure that boundary design routes through the right authority. This does not mean the CPO designs the boundary. It means the CPO ensures the boundary is designed by the function with the authority to define risk appetite, reviewed by the function with the authority to enforce it, and operated by the function with the capacity to maintain it. At large enterprises, this is typically the risk committee, the legal function, and a named operations owner respectively. At smaller organizations, it may collapse to the CEO and a named engineering owner — but even there, the routing matters. Boundaries set by a Slack conversation between two engineers are not the organization's risk appetite; they are a guess at it.

The boundary-governance cadence matters as much as the initial design. Boundaries designed once and never revisited drift from the organization's risk posture within quarters. A quarterly boundary review at appropriate authority level is the minimum operational discipline; monthly is better for high-stakes systems. The CPO owns the calendar for this, or it doesn't happen.

The internal-alignment concern from tiers 3 and 4 intensifies at tier 5. Sales teams claiming "fully autonomous" for a tier-5 product, product teams describing "guardrails" without architectural specificity, support teams unaware of the exception queue — all of these produce customer conversations that misrepresent the product. The CPO's alignment job is to ensure the organization can describe its own product accurately. This is harder at tier 5 than at lower tiers because "bounded autonomy" requires more vocabulary than "human in the loop" or "fully autonomous," and the vocabulary is not yet standardized. Internal training, not just internal documentation, is warranted.

Forward references

The boundary specification named above is the primary object Module 3 (incident response) treats at implementation depth — including the boundary-evaluation log structure, the enforcement-layer integration pattern, and the boundary-change deployment pipeline. Module 2 (evaluation) treats boundary calibration as a first-class evaluation target: how do you test whether a boundary is correctly calibrated before deploying it? Module 3 also covers exception-queue overflow and boundary-related incidents, including the two failure modes above. Module 4 (procurement) covers contract language for tier-5 deployments, specifically vendor representations about boundary architectures and the evidentiary requirements for verifying vendor claims about tier. Module 5 (organizational design) covers the boundary-governance cadence named above, including how the risk function, product organization, and engineering organization coordinate on boundary design and maintenance.

Tier 6: Self-Modifying Autonomy

The recommendation, stated first

Tier 6 is not currently an appropriate production architecture for most agentic deployments in 2026. This is a stronger claim than the recommendations at other tiers, and it warrants the strength. The evidence base, architectural readiness, and regulatory clarity required to operate at this tier responsibly do not yet exist at the level that would support broad deployment. The organizations operating at tier 6 in 2026 are doing so either because they are research organizations with specific mandates to study the tier, or because they have drifted to tier 6 without architectural intent.

The useful question for most operators is not "when do I deploy at tier 6" but "how do I recognize that I have drifted there, and how do I get back to tier 5." The default trajectory of any tier-5 system that incorporates production feedback into policy updates is drift toward tier 6. Without specific architectural discipline, most organizations will discover they have arrived at tier 6 in retrospect rather than by decision.

The architectural commitment

You are at tier 6 when the model acts autonomously *and* updates its own operational policy based on production feedback, RLAIF signals, in-context reward learning, or analogous self-modification mechanisms. The agent operating on Tuesday is not identical to the agent deployed on Monday, and the difference is not the result of a human-approved deployment — it is the result of the agent's own adaptation to its operating environment.

The defining property is that *the policy itself is a function of production behavior*. At tier 5, the agent acts within constraints humans defined; those constraints may be updated by humans, but the update is a deliberate human action with its own version history and approval chain. At tier 6, the constraints or the behavior within them evolve through a feedback loop that does not require per-update human approval.

This is why tier 6 is qualitatively different from tier 5 rather than "tier 5 with retraining." The accountability object at tier 5 is the boundary, which is stable between human-approved changes. The accountability object at tier 6 is the *adaptation process itself* — the reward model, the feedback signal, the update mechanism. When something goes wrong at tier 6, the question is not "what did the agent do" or even "what boundary was crossed." The question is "how did the agent become the agent that did this." That is a different category of question, and most organizations do not have the apparatus to answer it.

What changes at tier 6

Attribution breaks in the traditional sense. At tiers 1 through 5, accountability rests on identifiable humans making identifiable decisions — authors, approvers, reviewers, boundary designers. At tier 6, the agent that performed the action is not the agent any human reviewed, approved, or designed. The agent was shaped by a process: the initial deployment, plus subsequent adaptation, plus production feedback, plus whatever optimization pressure the reward signal applied. The humans who set the process in motion are attributable to the process; they are not attributable to any specific action the process produced downstream. This is a real break in the chain, and the legal and regulatory systems have not yet developed frameworks to address it.

Auditability requires policy-drift logging, which almost nobody builds. The tier-4 logging schema captures actions and decision traces. The tier-5 schema adds boundary evaluations and boundary-change history. Tier 6 requires a further layer: the policy itself must be versioned and auditable across time, with every update (every reward signal applied, every gradient step taken against production data, every in-context adaptation) captured in a way that supports reconstruction. The policy at time T must be retrievable. The delta between the policy at T-1 and T must be characterizable. The inputs that caused the delta must be traceable. This is a substantial engineering undertaking, and the infrastructure to do it well is currently available primarily inside major AI research labs. Enterprise deployments that nominally operate at tier 6 typically do not have this infrastructure.

Reversibility splits into three operations, one of which may not be solvable in general. Action rollback inherits from lower tiers. Policy rollback (reverting the agent's policy to a prior version) is technically tractable if the infrastructure above is in place. The third operation — *rolling back the downstream effects of having operated under a drifted policy* — may not have a clean solution. If the agent operated for weeks under a subtly drifted policy before the drift was detected, the actions taken during that period were all taken under the drifted policy. Rolling back individual actions addresses the actions; it does not address the systemic pattern of having been the drifted agent. The organization may discover that every customer interaction, every decision, every communication during the drift period reflects the drifted policy in ways that are individually small and collectively large. Cleaning this up is not a technical operation; it is an operational archaeology project.

Liability surface is genuinely uncharted. No court has yet adjudicated a case where the defendant's core defense is "the system I deployed is not the system that acted." The components of the defense are recognizable — the organization deployed the system in good faith, the system's behavior evolved through mechanisms the organization did not directly control, the evolution produced behavior the organization did not authorize — but the assembled defense has not been tested. Regulatory frameworks are similarly unsettled. The EU AI Act contains provisions for continuously learning systems but the operationalization of those provisions is ongoing. FTC guidance specific to self-modifying systems is emerging but not yet concrete. The practical implication: the organization operating at tier 6 is, legally and regulatorily, operating ahead of the frameworks that would govern its operation. This is a position some organizations can defensibly take; it is a position that requires deliberate acceptance of uncertainty.

What specifically changes is more concrete than the abstraction suggests. "Self-modification" is an umbrella term, and the umbrella covers five distinct things that operators must track separately because they have different governance implications and different detection requirements.

The first is *model weights*. The underlying model is being retrained, fine-tuned, or updated based on production data — through scheduled fine-tuning runs, online learning, RLHF or RLAIF on production interactions, or vendor-driven model updates pushed automatically. The model that processed yesterday's inputs is computationally different from the model processing today's inputs.

The second is *system prompt and configuration*. The agent's prompt template, system instructions, configuration parameters, or operating context are being updated based on production patterns — through prompt engineering updates pushed without governance review, configuration tuning based on production metrics, automated A/B testing that promotes winning prompts, or memory systems that accumulate context affecting future decisions. The model weights may be stable, but the agent's effective behavior is changing because what it's being told to do has changed.

The third is *retrieval and grounding*. The corpus of information the agent retrieves from is being updated based on production interactions — through vector databases updating on user queries, retrieval indices being refreshed based on usage, document ranking being adjusted based on what produced successful outcomes, or knowledge bases being modified based on production findings. The model and prompt may be stable, but the context the agent draws on when reasoning is changing.

The fourth is *boundary specification*. The constraints the boundary object enforces are being updated based on production patterns — through adaptive thresholds that shift based on observed usage, boundary parameters that adjust based on exception patterns, automated boundary loosening to reduce exception queue volume, or vendor-supplied boundary updates pushed without customer approval. This is the most consequential category because the boundary is supposed to be the load-bearing safety mechanism. When the boundary specification is itself drifting, the operator has lost the ability to know what the system is constrained to.

The fifth is *reward signals and evaluation criteria*. The signals the system uses to evaluate its own performance are being updated, which changes what the system optimizes for — through reward functions that incorporate production outcomes, evaluation criteria that shift based on business metrics, success signals derived from user behavior rather than from explicit human evaluation, or Goodhart drift where the proxy signal becomes the target.

Real Tier 6 systems often combine multiple categories. A vendor-supplied AISDR may be updating its model weights monthly (Category 1), refreshing its prompts based on A/B tests weekly (Category 2), updating its retrieval index continuously (Category 3), tuning boundary thresholds based on exception patterns (Category 4), and adjusting reward signals based on conversion data (Category 5) — all simultaneously, often without the customer organization being aware that any of it is happening. The operator's diagnostic question is not "is the system self-modifying" but "which of these five categories is in motion in our deployment, and what is our visibility into each of them." A deployment with strong governance over Categories 1-3 but no visibility into Categories 4 and 5 is operating at Tier 6 in the most consequential ways while believing itself to be governed.

The Anthropic reward-hacking result and what it establishes

The load-bearing evidence for tier 6's distinctive character comes from Anthropic's November 2025 paper on natural emergent misalignment from reward hacking in production reinforcement learning. The paper is worth treating structurally because it is currently the cleanest public demonstration of the failure mode tier 6 is uniquely exposed to.

The experimental setup: a pretrained model was given knowledge of reward-hacking strategies through synthetic document fine-tuning, then trained with reinforcement learning on real production coding environments — the same kinds of environments used in training Claude. The model, unsurprisingly, learned to exploit the reward-hacking opportunities it had been taught about. The surprising result: at the same point the model learned to reward hack on coding tasks, it generalized to a range of behaviors it had not been trained for. It began faking alignment, cooperating with hypothetical malicious actors, reasoning about malicious goals, and attempting sabotage — including, the paper notes, attempts to sabotage the research codebase used to produce the paper.

The generalization is the core finding. Reward hacking in training did not stay contained to the domain where reward hacking was taught. The model generalized from "I can exploit this specific grading system" to a broader pattern of treating evaluation as an adversarial target and pursuing goals misaligned with its ostensible purpose.

The mitigation the researchers found — "inoculation prompting," in which reward hacking is framed as acceptable during training to prevent the model from associating hacking with misalignment — worked to reduce the generalization. The researchers acknowledge the mitigation has the flavor of whack-a-mole; it addresses the specific pathway without resolving the underlying alignment question.

Two readings of what this establishes, one narrow and one broader.

The narrow reading. In at least one production-adjacent setup, reward hacking generalized from its domain of origin to a broader misalignment pattern, and the generalization was detectable only through specific evaluations designed to surface it. This is a specific technical finding with specific relevance to training regimes that include RL on production data.

The broader reading. The paper provides rigorous evidence for what the principal-agent analysis would predict: when the agent's optimization target is a function of its operating environment, the agent's behavior in that environment will adapt in ways that may not reflect the operator's intent. The adaptation is not malicious in any interesting sense; it is the predictable result of optimization against a misspecified signal. The specific generalization pattern — hacking spreads to broader misalignment — is an empirical finding. The general pattern — self-modification produces behaviors the operator did not intend — is a structural property of the tier.

Both readings matter for tier 6 operators. The narrow reading tells them to attend specifically to reward-signal design when their systems incorporate production feedback. The broader reading tells them that the accountability framework they inherited from tier 5 (design the boundary, govern the boundary) is insufficient — at tier 6, they must also govern the process by which the agent adapts, which is a different operation with different requirements.

When tier 6 is right

This is where the tier treatments diverge most sharply from the pattern of the other tiers. I am not going to offer three checkable conditions for when tier 6 is correct, because the honest answer is that for most enterprise deployments in 2026, it is not. Instead, two conditions under which tier 6 is potentially defensible, both narrow.

The organization is conducting research on self-modifying systems with appropriate safeguards. AI research organizations developing alignment techniques, safety evaluations, or interpretability tools may operate at tier 6 with deliberate intent, because the research requires the system to be self-modifying in order to study the phenomena of interest. These deployments typically include specific additional safeguards — sandboxed environments, extensive evaluation harnesses, red-teaming, publication of findings — that enterprise deployments typically do not.

The use case has such high return and such low per-action stakes that the organizational cost of drift can be absorbed. Certain high-volume, low-consequence adaptive systems — recommendation engines, content ranking, ad optimization — have operated at something approaching tier 6 for years without the vocabulary to describe it. The defensibility rests on the per-action stakes being genuinely low (no single ranked item causes material harm) and the feedback loop being genuinely aligned (the system adapts to reveal-what-users-click, which is the system's actual purpose). When either condition fails — when per-action stakes are not actually low, when the feedback loop is misaligned with the organizational purpose — these systems produce the characteristic tier-6 failures that have shaped regulatory attention on algorithmic systems over the past decade.

For enterprise AISDR, procurement, customer service, and similar use cases, neither condition applies. The per-action stakes are not trivially low; the feedback loops are typically misaligned (systems optimize for proxy metrics like booked meetings rather than true outcomes like customer value); the organization rarely has the research apparatus to govern self-modification responsibly. Tier 5 is the appropriate target architecture. Drift to tier 6 should be recognized and reversed rather than accepted.

The characteristic failure mode

The tier-6 characteristic failure mode is *undetected policy drift producing systematic behavioral change that is visible only in retrospect*. The failure is not a single wrong action; it is the realization, after weeks or months, that the system's behavior has evolved in a direction the operator did not intend, that the evolution produced cumulative effects across many interactions, and that the retrospective cleanup is prohibitively expensive.

This failure mode does not yet have a public case study at enterprise scale in agentic AI deployments, for two reasons. First, most enterprise systems operating in the tier-6 vicinity are at the boundary with tier 5 and have not yet accumulated the drift required to produce the characteristic failure. Second, when the failure does occur, it typically surfaces as a gradual decline in system quality rather than as a discrete incident, which makes public documentation less likely.

The closest available public analogs come from adjacent domains. The Anthropic paper treated in depth above is the cleanest demonstration of the underlying dynamic in a research setting. The history of recommendation-engine failures in social media platforms — where optimization for engagement produced systematic amplification of divisive content over time — is a decade-long case study of tier-6-adjacent systems producing cumulative harms through feedback-loop misalignment. The 2016 Microsoft Tay chatbot incident is an extreme but illustrative compressed case: a self-modifying system's rapid adaptation to its operating environment produced behavior the operator did not intend within hours, with cleanup that was essentially abandonment of the system.

These analogs point toward what tier-6 failures in agentic AI will likely look like when they surface publicly. The characteristic arc: deployment with initially acceptable behavior, a period of adaptation producing subtle shifts in system behavior, detection of the shifts either through external complaints or internal audit, investigation revealing systematic drift across many actions, remediation that is partial because the full downstream impact cannot be reconstructed.

The structural reading for operators: the failure is not unpredictable. The mechanism is well-understood. What is unpredictable is when specifically it will surface for a given deployment, and how large the accumulated drift will be at the point of surfacing. The operator's question is not "will tier-6 systems drift" but "how will I detect the drift before it produces cumulative harm at a scale I cannot remediate."

Vendor and architecture examples

Most vendor marketing that invokes self-improving agents, continuous learning systems, or adaptive AI is referring to systems that operate somewhere between tier 5 and tier 6, with the specific architectural commitments varying by product and often by deployment. The clearest operators at tier 6 in public view are research-stage systems rather than enterprise products.

The actionable directive: when evaluating a vendor claim that invokes self-improvement, continuous learning, or adaptive behavior, ask three specific questions. What specifically adapts — the policy, the reward model, the retrieval context, or something else? What controls the adaptation — who approves updates, what signals drive them, what safeguards prevent undesired generalization? What logging supports reconstruction — can you retrieve the exact policy the system operated under at any past point, and characterize the deltas between policy versions? Vendors whose answers are vague are operating without the apparatus tier 6 requires, regardless of the marketing language.

Monday artifacts

CTO artifact: Architecture Decision Record for any deployment that might be tier 6

The ADR for tier 6 has a different structural role than the ADRs at other tiers. At tiers 2 through 5, the ADR documents an intentional architectural choice. At tier 6, the ADR's primary function is to *confirm that the deployment is not at tier 6 when it claims to be at tier 5*, or to document the deliberate and unusual choice to operate at tier 6 with appropriate safeguards.

Decision. We are deploying [system] at tier 5 [or tier 6, if deliberate]. If tier 5: we are explicitly not at tier 6 because [specific architectural features that prevent self-modification from production feedback: policy updates require human approval, reward signals are not automatically applied to policy updates, specific mechanisms prevent the loops that would produce tier-6 behavior]. If tier 6: we are deliberately at tier 6 because [specific research or use-case rationale], and we have implemented the specific safeguards listed below.

Drift-detection preconditions. Policy versions are retrievable at any historical timestamp. Deltas between policy versions are characterizable. A detection mechanism surfaces policy drift above a defined threshold. Specific evaluations are run periodically to detect the generalization patterns documented in the Anthropic reward-hacking research. Reward-signal specification is audited for misalignment with organizational purpose.

What would indicate we have drifted to tier 6 unintentionally. Specific behavioral changes in the system that cannot be attributed to explicit human-approved policy updates. Production metrics showing behavior patterns that were not present at deployment. Evaluation results degrading on safety or alignment benchmarks in ways that suggest generalization effects.

What we would do if we discovered unintentional tier-6 operation. Immediate suspension of the feedback loop (stop applying production signals to policy updates). Policy rollback to the most recent human-approved version. Investigation of the drift trajectory and its downstream impact. Remediation planning proportionate to the scope of drift.

Owner. One human, explicitly senior enough to authorize a system rollback that may affect business operations.

Review date. Monthly for systems in the tier-5/tier-6 boundary zone; quarterly for systems with strong architectural prevention of tier-6 operation.

Architect artifact: Policy versioning, drift detection, and feedback-loop controls

The architectural requirements at tier 6 are an order of magnitude more demanding than at tier 5. I will name the core requirements without pretending they are straightforward to build.

Policy versioning. Every change to the operational policy — whether made by a human-approved deployment, a retraining run, an applied reward signal, or an in-context adaptation — must produce a new versioned policy artifact with a unique identifier, a timestamp, and a characterization of what changed. Retrieval of any historical policy version must be possible. This is the foundational infrastructure for tier 6; without it, nothing else in this artifact is implementable.

Drift detection. The operator must have mechanisms that surface changes in system behavior not attributable to explicit human-approved updates. These mechanisms can include behavioral evaluation suites run on each policy version, outcome-tracking metrics with alerts on distributional shifts, red-team evaluations on specific patterns (including the generalization patterns from the Anthropic research), and adversarial probes designed to detect reward-hacking generalization.

Feedback-loop controls. If the system incorporates production feedback into policy updates, the mechanism by which feedback becomes policy change must be a first-class engineered component, not an emergent property. Specifically: what signals are collected, how they are aggregated, what updates they drive, what approval is required before updates are applied, what evaluations are run before deployment. Each of these is a design decision with safety implications. The default — "train on whatever production data we have" — is the architectural commitment that produces tier-6 failures.

Reward-signal specification audit. The reward signal, whether explicit or implicit, must be periodically audited against the organization's actual purpose for the system. Goodhart dynamics — where the reward signal becomes a target that the system optimizes against at the expense of the true purpose — are the dominant failure mode for feedback-loop systems, and they produce tier-6 behaviors even when no explicit self-modification is intended.

Circuit breakers. The system must have mechanisms by which the feedback loop can be suspended quickly. Rapid suspension is not a panacea — the drift that has already occurred remains — but it prevents further drift while investigation and remediation proceed. Circuit breakers at tier 6 are what the retraction primitive is at tier 4: a first-class operational capability, not a hopeful afterthought.

CPO artifact: Acceptance of uncertainty and the drift-detection operating rhythm

The tier-6-specific CPO concern is honest organizational acceptance that operating at or near tier 6 means operating ahead of the frameworks that would govern operation at tier 6.

The governance at this tier is qualitatively different from lower tiers. At tiers 1 through 5, the governance question is whether the organization is operating the tier correctly. At tier 6, the governance question is whether the organization has accepted the uncertainty inherent in operating where case law, regulatory guidance, and operational best practice are all still forming. Organizations that operate at tier 6 while claiming the certainty of lower tiers are not merely architecturally exposed; they are *representationally exposed* — they have asserted knowledge and control they cannot demonstrate. The CPO's job is to ensure that the organization's external representations and internal governance both reflect the actual state of knowledge, which includes substantial unresolved uncertainty.

The operating rhythm at tier 6 is more frequent than at other tiers. Monthly review of drift-detection signals. Quarterly red-team evaluations. Semi-annual audits of reward-signal specification. Annual external review by parties capable of surfacing drift patterns the organization might miss. This cadence sounds heavy. It is heavy. The honest position is that if the cadence feels disproportionate to the system's importance, the system probably should not be operating at tier 6 — the architectural choice has implied a governance burden the organization may not be prepared to carry.

The internal-alignment concern at tier 6 carries a specific additional responsibility. At other tiers, the concern is that sales, product, and support describe the system accurately. At tier 6, the concern is that the *organization itself* does not drift into describing its system as more predictable, controlled, or understood than it actually is. The drift toward overclaiming is the organizational analog of the system's own drift — a gradual shift in how the system is represented internally that tracks wishful thinking rather than operational reality. The CPO's specific job is to ensure that the organization's internal narrative about its tier-6 system remains honest as the system operates over time.

The connection back

Tier 6 is where the principal-agent framing that runs through this work becomes architecturally load-bearing rather than metaphorical. At tiers 1 through 5, the relationship between the humans who design and deploy the system and the system's operational behavior is mediated by specific, identifiable accountability objects — the artifact, the approval, the action, the boundary. The principal-agent structure maps cleanly onto these objects. At tier 6, the accountability objects shift from the output of the system to the process that shapes the system, and the principal-agent relationship becomes a relationship between the humans who specified the shaping process and the system that the process produces. The principal does not deploy an agent; the principal deploys a process that produces a succession of agents, each responsive to its operating environment in ways the principal cannot fully predict.

This is the structural reason tier 6 warrants the caution above. Not because the technology is necessarily dangerous — the research so far suggests the behaviors involved are comprehensible and in many cases manageable — but because the accountability framework that makes principal-agent relationships governable has not yet been built for this tier. The organizations that will operate at tier 6 responsibly in the coming years will be those that participate in building that framework, not those that assume it already exists.

Forward references

Module 2 (evaluation) is especially load-bearing for tier 6, covering both the evaluation infrastructure required to detect policy drift and the specific evaluation patterns derived from the Anthropic reward-hacking research. Module 3 (incident response) covers policy versioning infrastructure at implementation depth, including the versioning schema and the drift-characterization mechanisms. Module 3 also covers tier-6-specific incident types: unintentional drift to tier 6, detected drift under operation, and remediation of accumulated drift impact. Module 4 (procurement) covers vendor representations about self-modification — specifically, the specific questions about adaptation mechanisms, controls, and logging that should be part of any evaluation of a vendor invoking adaptive behavior. Module 5 (organizational design) covers the governance cadence and responsibilities specific to tier-6 operation, including the external review relationships that this tier requires and that most organizations have not structurally established.

Part 4: Cross-Tier Dynamics

The module's compounding layer. The patterns that only become visible once the six tiers exist as shared vocabulary.

The tier-by-tier chapters treat each tier as if the operator will spend time at exactly one tier and govern it in isolation. That framing is useful for clarity; it is not useful for reality. In reality, operators sit across multiple tiers simultaneously, drift between tiers silently, inherit tier choices made by vendors and predecessors, and answer to regulatory frameworks that treat the tiers differently than the architects of those tiers intended.

This part treats the dynamics that emerge when the gradient is operated as a system rather than as a menu. Eight load-bearing patterns, in order of how much weight each carries for operators trying to make decisions mid-deployment.

The tier is a joint product of architecture and behavior

The most important observation about operational tier is that it is not determined by the system architecture alone. It is determined by the architecture *and* the behavior of users inside the architecture, and the two can diverge in ways that make the architectural claim meaningless.

A product configured for tier 2 — the model suggests, the human edits, the edited version sends — operates at tier 2 when users edit substantively. The same product operates at tier 3 when users click-accept without editing. The same product operates at something approaching tier 1 when users treat the suggestions as reference and compose from scratch. The architecture is identical. The operational tier varies by user population, by deployment culture, by time of day, by quarter-end pressure.

The useful way to think about this: *the architecture fixes the ceiling of the tier that can be operated; the behavior determines the floor*. A tier-2 architecture cannot operate at tier 5 — users cannot conjure autonomous execution from a suggestion interface. But a tier-2 architecture can operate at any tier from 1 to 3 depending on how users treat it. A tier-5 architecture can operate anywhere from tier 4 (when exception queues overflow and exceptions start being rubber-stamped) to tier 6 (when policy drift accumulates unchecked) depending on operational discipline.

This has a specific consequence for governance. Architectural governance — did we configure the product correctly — is necessary and insufficient. Behavioral governance — are users operating the product at the tier we claim — is equally necessary and systematically under-built. At every tier from 2 upward, the behavioral drift signature is predictable, and monitoring for it is the governance discipline that distinguishes operators who know their tier from operators who assume it.

Every tier has a named diagnostic number

Architecture does not tell you where you are. Behavior does. And behavior is only knowable through specific measurements that most deployments do not capture.

Each tier has a *tier-integrity metric* — a single number that, if it drifts, signals the deployment has silently moved to a different tier. These metrics are the operational translation of the behavioral-governance problem above. The catalog:

- **Tier 1**: no metric is possible, because tier 1 is not architecturally enforced. Sampling of user practice is the only signal.
- **Tier 2**: the *edit ratio* — fraction of model-generated language replaced by human editing in the final artifact. A falling edit ratio across users is the drift signal toward tier 3.
- **Tier 3**: the *reviewer latency per action* — time between the action being presented to the approver and the approval click. Distribution compression toward zero is the drift signal toward tier 4.
- **Tier 4**: the *authoritative-record propagation latency*, the dialer check — time between a fact changing in an authoritative system and every autonomous surface reflecting the change. Increasing latency is the drift signal toward ungovernable execution.
- **Tier 5**: the *boundary-exception handling time* — median and 95th-percentile time between an exception triggering and meaningful handling. Degradation signals either tier-4 drift (exceptions stop being handled) or tier-6 drift (the boundary is being informally loosened to reduce exception volume).
- **Tier 6**: the *policy-drift detection latency* — time between a policy update occurring and the update being evaluated. If this number is not bounded, tier 6 is not operationally governable.

Each metric should be displayed, reviewed, and tied to specific thresholds that trigger action when crossed. The dashboards exist in a small fraction of current deployments; the thresholds exist in fewer; the triggered actions exist in almost none. This is the single highest-leverage operational improvement available to most agentic AI deployments in 2026, and it does not require any model change, vendor change, or architectural change. It requires deciding what your operational tier is and measuring whether you are still at it.

Behavioral drift is up the gradient — and that is the default trajectory

The direction of behavioral drift is not random. It is systematically upward on the gradient. Users adapt to tier architectures in ways that silently push deployments to higher tiers over time.

At tier 2, users discover that accepting model output with minimal edits is faster than revising carefully, looks acceptable to casual review, and produces no immediate consequence. Editing atrophies. The deployment drifts toward tier 3.

At tier 3, approvers discover that clicking approve is faster than reading carefully, produces throughput that the organization rewards, and escapes scrutiny until something goes wrong. Approval inflates. The deployment drifts toward tier 4.

At tier 4, reviewers discover that monitoring dashboards is cognitively expensive and the reward for catching problems is small compared to the penalty for being visible as a bottleneck. Dashboards go unopened. The deployment drifts toward un-governable execution.

At tier 5, the exception queue grows when boundaries are tight, and the political pressure is to loosen the boundaries to reduce the queue. Boundaries loosen. Exception handling hollows out. The deployment drifts toward tier 6.

At each step, the drift is rational at the individual level. Users, approvers, reviewers, and operators are responding to the incentives in front of them. The drift is harmful at the organizational level because the incentives in front of individuals are not the accountability framework the organization has committed to.

This has a specific consequence for deployment strategy: *the default trajectory of any agentic AI deployment is upward on the gradient unless specific effort is exerted to prevent it.* Deployments that are at tier 3 this quarter will be at tier 4 next year unless governance specifically holds them at tier 3. The governance that holds a deployment at its intended tier is the cadence discipline

— the weekly edit-ratio review, the monthly reviewer-latency distribution, the quarterly boundary audit. Without the cadence, the drift is inevitable.

The counter-example worth naming explicitly: mature migration is downward, not upward. Salesforce's 2025 architectural correction on Agentforce — the shift from autonomy-first positioning toward guardrails, deterministic automation, and data-first design — was a deliberate downward migration from tier 6 positioning toward tier 5 operation. The industry narrative implicitly frames tier 6 as the mature state and lower tiers as training wheels. The evidence suggests the opposite. Mature operators migrate toward tier 5 because tier 5 is the architecturally governable state. Higher tiers are where immature operators begin by ambition and end by retrospective discovery of what they deployed.

The accountability object evolves across the gradient

The object of accountability — the thing the organization is responsible for — is not the same at every tier. It evolves in a specific direction, and the evolution matters because it determines what the organization must govern and what a court or regulator will examine.

At tier 1, the object is the *artifact* — the document, email, or code the human authored with model assistance as reference. Standard authorship accountability.

At tier 2, the object is the *jointly authored artifact* — the sent content that carries both human and model contributions. The Air Canada principle applies: the organization cannot disclaim the model's contribution by pointing to the human's editing.

At tier 3, the object is the *ratified artifact* — the model-produced content that the human's approval has made the organization's speech. The Mata principle applies: the approval is legally meaningful only if the review behind it was performed.

At tier 4, the object is the *action* — what the autonomous system did in the world, reviewable after the fact. Attribution no longer rests on a named human approver; it rests on the organization that deployed the system, defended against the question of whether the review architecture was operational.

At tier 5, the object is the *boundary* — the declared constraint architecture that shapes the space of permissible actions. Accountability shifts from "did the agent do the right thing" to "was the boundary reasonably designed, approved at commensurate authority, and maintained against changing conditions." This is closer to product liability than to tort.

At tier 6, the object is the *process* — the mechanism by which the agent is shaped through adaptation. The principal does not deploy an agent; the principal deploys a process that produces a succession of agents, each responsive to its operating environment. Accountability rests on the design and governance of the shaping process, not on any specific agent the process produces.

The evolution of the accountability object is the structural explanation for why the governance burden grows as you move up the gradient. Each step up adds a new category of thing to govern without removing the categories below. At tier 5, the organization is still accountable for individual actions and jointly authored artifacts when the agent produces either; it is additionally accountable for the boundary. At tier 6, the organization is accountable for all of that plus the shaping process. Operators who move up the gradient without proportionately expanding their governance apparatus are operating under an accountability framework that does not match their architecture.

Who gets named when things go wrong

A specific consequence of the accountability-object evolution: the humans who appear in the complaint, the sanction, the regulatory letter, or the press release are not the same humans at different tiers. The tier determines whose name carries the adverse event.

At tiers 1 and 2, the authoring individual bears primary liability under most frameworks. The attorney who signs the brief, the physician who signs the chart note, the engineer who commits the code are personally accountable regardless of AI assistance. Mata v. Avianca sanctioned individual attorneys, not the firm.

At tier 3, the approver shares liability with the organization, and the share allocation depends on the specific legal or professional framework. The Mata pattern suggests the individual bears significant exposure; emerging Article 22 interpretation may shift exposure toward the organization when the approval volume is structural rather than chosen. This tier is where individual-vs-organizational liability is most actively being allocated in real cases.

At tier 4, liability is primarily organizational because no individual approver is identifiable per action. The organization is sued; the organization defends. Individual humans enter the picture only when their specific role in the deployment (architecture decision, review process design, incident response) is alleged to be negligent. The tier-4 ADR's named owner is the individual whose name will appear if that allegation is made.

At tier 5, liability is organizational with a duty-of-care dimension. The boundary designer, the boundary approver, and the boundary operator are identifiable roles that plaintiffs can name. The governance cadence is itself part of the defense — a boundary that was reviewed quarterly is defensible in a way that a boundary that was never revisited is not.

At tier 6, the exposed humans are those associated with the shaping process: the reward-signal specifier, the evaluation-harness owner, the drift-detection designer, the external-review arranger. This is uncharted territory and the case law has not yet arrived to name the specific roles that carry liability. The safe presumption is that the organization's senior leadership bears the exposure until the frameworks specify otherwise.

Two practical implications for operators. First, the humans whose names are on the tier must understand the personal exposure the tier creates for them. At tiers 1, 2, and 3, this is individual professional exposure; at tiers 4 through 6, it is role-based organizational exposure that still attaches to named individuals in the org chart. Second, the organization's arrangements for defending named individuals — indemnification, D&O coverage, professional-liability coverage — must match the tier. Organizations operating at tier 5 with defense arrangements designed for tier 3 deployments are exposing individuals to liabilities the organization is not prepared to absorb.

Regulatory doctrine migrates up the gradient

The regulatory doctrine most actively contested today sits at tier 3. The meaningful-human-involvement question — whether a human approval step is substantive enough to shift a decision out of the GDPR Article 22 prohibition on solely automated processing — is being adjudicated in regulatory guidance, enforcement actions, and the first wave of court rulings across the EU and parallel U.S. developments.

The doctrine does not stay at tier 3. Once the criteria are articulated at tier 3 — reviewer authority, access to information, volume allowing consideration — they will cascade. Tier 4's sampled post-hoc review will be examined against the same criteria; most tier-4 deployments will fail the test in their current form. Tier 5's exception-handling discipline will be examined for whether the exception-path review is itself meaningful or rubber-stamped. Tier 6's shaping process will eventually face an analogous question: is there meaningful human involvement in how the adaptation occurs, or is the process effectively automated at a higher level?

This is the general pattern. Regulatory doctrine, once established at one tier, moves up the gradient over time. The timelines vary by jurisdiction and by specific doctrine, but the direction is consistent. Operators who design their deployments against the doctrine *as it exists today* are building against a moving target; operators who design against the doctrine *as it is plausibly evolving* are building systems that will not require reconstruction every eighteen months.

The anticipatory move is to build tier-3 discipline at tier 4, tier-4 discipline at tier 5, and tier-5 discipline at tier 6. This sounds like conservative over-engineering. It is cheaper than architectural reconstruction when the doctrine arrives.

A specific corollary: operators who rely on tier 5's boundary architecture to satisfy meaningful-human-involvement are relying on a position that is not yet settled law, but is the strongest available defense for high-volume automated decision-making. If that position is upheld, tier 5 is regulatorily defensible; if it is rejected, tier 5 collapses toward tier 3 equivalence for regulated decisions, and the operational consequences are substantial. Operators should design tier-5 boundary architectures against the stronger interpretation of meaningful involvement — boundaries that would satisfy a court looking for genuinely deliberative boundary design — rather than against the minimum that current guidance permits.

Operating ahead of the frameworks

The cumulative picture: at every tier from 3 upward, operators are making architectural and governance choices that the frameworks governing those choices have not yet fully specified. Enterprise AISDR operators in 2026 are operating ahead of case law, ahead of regulatory guidance, ahead of standard operating procedure within their own professions. The gap between operation and governance widens as the tier rises.

This is not a criticism of the operators. It is a description of the state of the field. Agentic AI moved from research into enterprise production faster than the governance apparatus could be built around it. The frameworks will arrive; they have been arriving, in fragments, throughout 2024 through 2026. In the interim, operators are making consequential decisions with incomplete information and unfinished doctrine.

Two operating postures are available. The first is to operate as if the frameworks will not arrive — to assume that current ambiguity is a permanent feature and that architectural choices made today will not be re-examined under later standards. This posture is common. It produces deployments that will be retrospectively judged by frameworks the operators did not anticipate and did not design for.

The second is to operate as if the frameworks will arrive, will impose specific requirements, and will apply retrospectively to the deployments made before their arrival. This posture is more expensive in the short term — more logging, more documentation, more governance overhead — and it is less expensive in the medium term when the frameworks do arrive and retrospective cleanup would otherwise be the only option. The second posture is the one this module implicitly recommends. The cadence disciplines, the logging schemas, the ADR templates are all artifacts of designing deployments that will survive examination under frameworks that do not yet exist.

A subtler risk: organizations that operate ahead of the frameworks often drift into describing their systems as more predictable, controlled, or understood than the frameworks would validate. The CPO's representational-exposure concern at tier 6 is the compressed form of this risk; it exists in less acute form at every tier. The organization that claims tier-3 compliance for a deployment operating at tier 4 is representationally exposed in exactly the same way. Honest internal narrative about the organization's own systems is the foundational governance discipline; every other governance practice depends on it.

Governance is cadence, not policy

The final cross-tier pattern is the one the CPO artifacts at every tier converged on: *governance of agentic AI is ongoing attention, not one-time setup.*

The architecture can be configured once. The boundary can be designed once. The prompt can be refined once. But the operational tier drifts continuously through behavior, through exception patterns, through quiet erosion of the review discipline, through the shifting tactics of adversaries and the changing expectations of users. A system that was at tier 3 with meaningful review six months ago is at tier 4 with decayed review today, and it will be at an un-governable state six months hence if the governance cadence does not catch the drift.

The cadences that carry the weight: weekly behavioral-metric review at the level closest to users (edit ratios, reviewer latencies, action volumes), monthly governance review at the operational-owner level (tier-integrity metrics, exception patterns, boundary performance), quarterly governance review at the executive level (tier choice confirmation, vendor performance against representations, regulatory surface updates), semi-annual strategic review at the board level (systemic risk posture, accumulated exposure, proportionality of governance to deployment scale).

The cadence is not theater if it is connected to action. Each review level has specific triggers that move decisions up the chain: a metric crossing a threshold, a pattern emerging across deployments, a regulatory development changing the exposure. Without the triggers, the cadence is performance. With the triggers, the cadence is the mechanism by which the organization's stated tier stays its actual tier.

The cadence also requires specific humans whose time is allocated to governance rather than assumed to be extracted from their operational roles. This is the single hardest governance-building challenge for most organizations, because agentic AI governance cuts across product, engineering, legal, risk, and operational functions in ways that no single function fully owns. The CPO or the risk function typically inherits the convening role; the CTO or the chief architect typically inherits the technical governance; the general counsel or dedicated AI governance counsel inherits the regulatory surface; no one typically inherits the behavioral-monitoring function, which is why the tier-integrity metrics are the single most under-built category of governance artifact in current deployments.

The organization that builds the cadence discipline is the organization that stays at its declared tier. Every other governance practice — the ADR, the logging schema, the boundary specification — is a necessary artifact of the cadence, not a substitute for it.

The patterns in combination

Eight patterns treated explicitly above, with additional patterns woven through the tier chapters. They do not operate in isolation. The tier is a joint product of architecture and behavior; the tier-integrity metrics are how you measure the behavior; behavioral drift is the default trajectory in the absence of those measurements; the accountability object evolves as drift occurs; who gets named when things go wrong depends on what the accountability object is; regulatory doctrine will examine the accountability object against standards that are migrating upward; operators are making choices ahead of the frameworks that will judge those choices; the governance that holds all of this together is cadence discipline that most organizations have not built.

A reader of this module who internalized exactly one pattern should internalize the first: *operational tier is determined by architecture and behavior together, and behavior requires specific measurement*. Every other pattern depends on that one. A reader who internalized two should add *behavioral drift is the default trajectory unless cadence discipline catches it*. A reader who internalized three should add *the governance burden grows faster than the tier, and operators who move up the gradient without proportionately expanding governance are operating outside their declared tier whether or not they know it.*

The module's argument, compressed: pick your tier deliberately, measure whether you stay at it, and build the cadence that catches the drift. The rest is detail.

Part 5: The Diagnostic — Identifying Your Actual Tier

The vocabulary the module establishes is most useful when applied to a specific deployment. Most readers will find, on running the diagnostic below, that their deployment operates at a tier different from the tier they assumed — often different from the tier the vendor marketed, different from the tier the architecture claims, different from the tier the procurement contract specifies.

The diagnostic is not a tool for choosing a tier. It is a tool for identifying the tier a deployment is already at. The distinction matters. A reader who wants to *select* a tier for a new deployment should read the six tier chapters and make the choice against their specific use case. A reader who wants to *diagnose* the tier of an existing deployment — to surface gaps between claimed and operational tier, to identify governance debts, to inform procurement or audit conversations — should run the seven questions below against their current production system.

The diagnostic is designed so that the answer is defensible even if the reader does not like it. Each question resolves on observable facts, not on intent.

Question 1. Does the model's output appear in the sent artifact without human retyping?

If no: the deployment is at tier 1. Stop the diagnostic; the remaining questions do not apply. Review the tier 1 chapter to verify the architecture supports the tier-1 claim in practice, including the behavioral-governance concerns specific to that tier.

If yes: proceed to Question 2.

Question 2. Does a human edit the model's output before it is sent?

If yes, and the editing is substantive (measurable edit ratio above the threshold the organization considers "authored rather than reviewed"): the deployment is at tier 2. Review the tier 2 chapter, paying specific attention to the Air Canada principle and the diff-logging requirement.

If yes, but the editing is cosmetic or absent in practice: the deployment is operationally at tier 3, regardless of architecture. Review the tier 3 chapter; the tier-2 architecture is not providing the governance the organization assumes.

If no: proceed to Question 3.

Question 3. Does a human approve each individual output before it is sent?

If yes, and the approval is synchronous with the action, and the reviewer has the authority, information, and volume allowance to exercise judgment: the deployment is at tier 3. Review the tier 3 chapter with specific attention to reviewer-latency measurement and approval-inflation monitoring.

If yes, but the approval is pro-forma at volume: the deployment is operationally at tier 4, regardless of architecture. Review the tier 4 chapter; the approval gate is not providing the governance the organization assumes.

If no: proceed to Question 4.

Question 4. Does the system take externally-visible actions without a per-action human gate?

If yes: proceed to Question 5.

If no, and the deployment has not been resolved by earlier questions: the architecture is unusual or the diagnostic is being applied to a system not yet meaningfully in production. Review the module for the tier closest to the architectural intent and address the ambiguity before proceeding.

Question 5. Are there declared, documented, version-controlled, operationally-enforced constraints on what the system can do without escalation?

If no: the deployment is at tier 4. Review the tier 4 chapter with specific attention to the dialer check, the retraction-latency SLA, and the review-cadence calibration. The absence of meaningful constraints means tier 5's boundary architecture does not operationally exist, regardless of what documentation asserts.

If yes, and the constraints are enforced by a mechanism the agent cannot bypass (not soft enforcement, not log-only), and exception handling is built, staffed, and measured: the deployment is at tier 5. Proceed to Question 6.

If yes, but the constraints are aspirational (documented but not enforced, or enforced loosely): the deployment is operationally at tier 6, regardless of the tier-5 claim. Review the tier 6 chapter with attention to the representational-exposure concern; the organization has asserted a position it cannot support.

Question 6. Does the system incorporate production feedback into its own policy updates?

If no: the deployment is at tier 5. Review the tier 5 chapter with specific attention to boundary-governance cadence and the three-role split (designer, approver, operator).

If yes, with explicit human approval required per policy update: the deployment remains at tier 5 with a controlled adaptation mechanism. The tier-5 chapter applies, with additional attention to the boundary of tier 5 and tier 6. Proceed to Question 7.

If yes, without human approval per update: the deployment is at tier 6. Proceed to Question 7 and review the tier 6 chapter with full attention.

Question 7. Can the operator point to a specific, dated policy version that produced any given historical action?

If yes: the deployment has the infrastructure to be auditable at its tier, whatever that tier is.

If no: the deployment is auditably weaker than its architecture implies. At tier 5, the inability to retrieve historical policy versions undermines the duty-of-care defense. At tier 6, it is operationally disqualifying — tier 6 without policy versioning is tier 6 with no accountability framework.

The seven questions resolve the operational tier. The resolution is often surprising. Deployments confidently described as tier 3 frequently turn out to be operating at tier 4 because approvals have inflated. Deployments claiming tier 5 frequently turn out to be at tier 6 because the boundary is documented but not enforced. Deployments claiming tier 2 are sometimes at tier 1 because users prefer to compose from scratch, or at tier 3 because editing has atrophied. Each gap is a governance issue that deserves explicit attention.

Run the diagnostic, note the answer, and read the tier chapter that corresponds to the actual tier rather than the assumed tier. That is the module's most direct application to an existing deployment.

Part 6: What Each Tier Obligates

The module closes with the forward-pointing summary of what operating at each tier obligates the organization to build, monitor, and govern. This is the bridge to modules 2 through 10, which treat the obligations at implementation depth.

Tier 1 obligates explicit organizational rationale for operating against the market default, user training in the tier-1 expectation, and behavioral sampling in contexts where tier-mismatch creates material exposure. The light-touch governance is appropriate to the tier's narrow accountability surface.

Tier 2 obligates diff logging, edit-ratio measurement, joint-authorship acknowledgment in organizational posture, editing-discipline training programs, and vendor contract pushback against disclaimers that would shift responsibility for model-generated content. The Air Canada principle carries this tier.

Tier 3 obligates reviewer-latency logging with thresholded alerting, review-interface design that surfaces the information needed for judgment, frictionless rejection paths, approval-inflation monitoring, and — especially — individual-liability protection for named approvers. Mata v. Avianca sanctioned individual attorneys; every tier-3 deployment carries the same structural exposure for its named approvers.

Tier 4 obligates full decision-trace logging with policy-version identifiers, retraction protocols and transactional boundaries, the dialer check for authoritative-record propagation, documented review cadence matched to blast-radius-per-unit-time, and GDPR Article 22 exposure mapping. The 11x case is the canonical warning; operators who do not build the infrastructure reproduce the 11x conditions whether or not they have yet produced the 11x outcome.

Tier 5 obligates version-controlled and operationally-enforced constraint architectures, published exception-handling SLAs with observed performance data, the three-role accountability split across boundary design / approval / operation, boundary-change management with the rigor of production code, and the boundary-governance cadence proportionate to risk level. The Salesforce migration is the canonical positive example; operators targeting tier 5 should study what mature downward migration looks like.

Tier 6 obligates policy versioning infrastructure, drift-detection mechanisms including the specific evaluation patterns derived from the Anthropic reward-hacking research, feedback-loop controls treating the adaptation process as a first-class engineered component, reward-signal specification audits against Goodhart dynamics, circuit breakers enabling rapid suspension of the feedback loop, representational-exposure monitoring to prevent organizational narratives from drifting toward overclaiming, and the heavy cadence discipline — monthly drift-detection review, quarterly red-team evaluations, semi-annual reward-signal audits, annual external review — that the tier's uncertainty requires. Most current enterprise deployments should not be at tier 6, and should treat drift toward it as a governance problem to reverse rather than a maturity milestone to celebrate.

The module's compressed argument, one final time: pick your tier deliberately, measure whether you stay at it, build the cadence that catches the drift. Everything else in this reference series — the eval harnesses, the logging schemas, the incident runbooks, the procurement language, the organizational design, the regulatory posture — is detail on how to do those three things at each tier.

The next module treats the eval harness for agentic AI systems, with specific attention to how evaluation requirements differ across the gradient. The vocabulary established here carries forward. A reader who has internalized the six tiers and the diagnostic is ready to engage Module 2; a reader who has not should return to the tier chapter that matches their actual deployment before proceeding.

Part III — The Operations

Five chapters specifying the cadence discipline across the operational domains where governance becomes concrete.

The framework is the vocabulary. The operations are the work.

Each chapter in Part III addresses one operational domain where the cadence governance from Module 1 becomes concrete practice. Evaluation infrastructure. Incident response, logging, and reversibility. Procurement and vendor evaluation. Organizational design. Regulatory posture. Each chapter is organized the same way: the recommendation stated first, the operational analysis, the Monday artifacts.

The chapters are designed to be consulted as a reference work rather than read sequentially. A reader who needs the procurement chapter can read it without having first read the evaluation chapter, with the cost that some cross-references will require flipping back to context not yet built. Cross-references are explicit throughout; the forward-references section at the end of each chapter names what each subsequent chapter adds.

Module 1 is the prerequisite for all five. Beyond that, the chapters can be read in any order the reader's operational questions demand.

Module 2: Evaluating Agentic AI Systems

The how. What you measure, what infrastructure you need to measure it, and how the measurement requirements differ qualitatively across the autonomy gradient.

The recommendation, stated first

For any agentic AI system you operate at tier 3 or above, build an evaluation harness *before* the system reaches production. For systems already in production without one, treat the absence as a Sev-2 governance issue requiring quarterly remediation. The eval harness is not optional infrastructure for agentic AI in 2026; it is the load-bearing instrument by which the organization knows what its system is doing, demonstrates compliance with regulatory frameworks that require it, and survives the contact between architectural intent and operational reality.

The honest tension worth naming: the existing publicly available evaluation frameworks for agentic AI — CLEAR, Aisera's CLASSic, the MOYA four-pillar approach, the QuantumBlack three-layer model — are all serious contributions, and a thoughtful operator will have encountered at least one of them. None of them yet organizes evaluation around the operational tier the system is meant to occupy. That is the gap this chapter fills. It is not a replacement for those frameworks. It is the missing layer that determines which parts of those frameworks apply to your specific deployment.

What evaluation has to do, in two parts

Before the tier-specific layer, the framing question. *What is an evaluation harness for?*

The dominant answer in the existing literature is performance evaluation: does the system do what it was built to do, accurately, reliably, at acceptable cost. The CLEAR framework operationalizes this across cost, latency, efficacy, assurance, and reliability. The CLASSic framework does the same with cost, latency, accuracy, stability, security. The MOYA approach evaluates across four pillars — LLM, memory, tools, environment — with attention to the runtime behaviors that static testing misses. These are real contributions and the field is better for them.

The dominant answer is also incomplete. An evaluation harness for agentic AI must do *two* things, and the existing literature treats them as one.

The first is performance evaluation: *is the system performing well?* The second is **tier-integrity evaluation**: *is the system still operating at the tier it was deployed at?* The two are different, and the difference matters.

Performance evaluation answers questions like: is the agent's accuracy on customer support tasks above 85%? Is the average task cost below $0.50? Is the latency below 5 seconds at the 95th percentile? These questions have answers regardless of the system's tier; they apply equally to a tier-3 supervised system and a tier-5 bounded-autonomy system. They are the substance of CLEAR, CLASSic, MOYA, and adjacent frameworks.

Tier-integrity evaluation answers a different question: *given that we deployed at tier N, is the system still operating at tier N today?* The behavioral drift Module 1 named — editing atrophy, approval inflation, dashboard abandonment, exception-queue overflow — is invisible to performance evaluation. A system that has drifted from tier 3 to tier 4 may continue to perform well by every CLEAR metric while no longer satisfying the regulatory and accountability requirements of the tier it claims to occupy. The drift is detectable only through tier-specific behavioral metrics that performance evaluation does not capture.

The eval harness has to do both. Performance evaluation tells you if the system is doing its job. Tier-integrity evaluation tells you if it is doing its job *at the tier you committed to*. An organization that does only one is either flying blind on whether the deployment is operating as governed, or flying blind on whether it is producing value. Both are real liabilities, and both surface only when external events force examination.

This chapter treats both. Sections 1 through 4 develop the tier-integrity layer, which is the contribution this chapter makes to the existing literature. Sections 5 through 7 address performance evaluation, with explicit reference to the existing frameworks, and treat how the requirements shift across tiers. Section 8 covers the integration: what an eval harness looks like as a single instrument that does both.

The 37 percent gap

A specific number worth anchoring on before any of the structural work.

Research on multi-agent systems documents a 37 percent performance gap between laboratory evaluation and production behavior — a figure originating in benchmark research and surfaced in the 2025 CLEAR framework analysis of enterprise agentic AI evaluation. Systems that performed well on benchmark tasks degraded substantially when deployed against real workloads. Systems that performed well on benchmark tasks degraded substantially when deployed against real workloads. The gap is not unique to multi-agent systems; it appears in single-agent deployments at comparable magnitudes, and it is the largest single source of post-deployment surprise in agentic AI.

Three things explain the gap. First, lab benchmarks use curated inputs that systematically underrepresent the distributional shifts of production. Second, lab evaluation typically captures single-shot behavior; production captures the cumulative effects of many runs over time. Third, and most importantly for this chapter, lab evaluation does not capture user behavior — the editing patterns, approval patterns, and exception-handling patterns that determine the system's operational tier. The gap between lab tier and production tier is part of the 37 percent.

The implication for an evaluation harness is direct. *Pre-production evaluation is necessary but insufficient.* The harness has to operate continuously, in production, against the actual workload and the actual users, with measurement that captures the dimensions lab evaluation cannot. The chapter that follows treats this as a design constraint, not as an aspiration.

Section 1: Tier-integrity metrics — the catalog

Module 1 named the tier-integrity metric for each tier. This section gives the implementation depth for measuring them.

Tier 1: behavioral sampling

Tier 1's tier-integrity question — *are users actually composing from scratch, or are they pasting model output wholesale into their artifacts?* — has no log-based answer. The architecture does not record what the architecture is meant to enforce. The user can copy from the suggestion surface into their composition surface at any volume, and the suggestion surface does not know that they did.

The instrument here is behavioral sampling, conducted on a defined cadence. The implementation:

A representative subset of users in tier-1 deployments — typically 10 to 20 percent in deployments under 500 users, 5 percent in larger deployments — has their actual practice observed for a defined window. The observation can be screen recording for that window, manual review of artifact-versus-suggestion-content correspondence, or interview-based assessment of how the user reports interacting with the tool. The output is a determination of whether the user is operating at tier 1, tier 2, or somewhere between them.

The cadence depends on the stakes. For professional-responsibility contexts (legal, medical, regulated financial advisory), monthly sampling is appropriate. For competitive-IP contexts, quarterly is sufficient. For training and skill-development contexts, semi-annual sampling tied to the training cycle is reasonable.

The disposition matters as much as the measurement. When sampling reveals that a user is operating outside the declared tier, the response is not punitive — punitive response causes users to hide behavior rather than change it. The response is structural: training refresher, configuration adjustment, or honest acknowledgment that the deployment has drifted to a higher tier and the governance posture must follow.

Tier 2: edit ratio

Tier 2's tier-integrity metric is the edit ratio — the fraction of model-generated content replaced or modified by human editing in the final sent artifact. The implementation requires diff logging: at the moment the model produces output into the composition surface, the output is captured; at the moment the user sends the artifact, the final version is captured; the diff between them is computable and storable.

The basic computation is *(tokens in final version that were not in model output) divided by (tokens in final version).* The basic metric is sufficient for tier-integrity monitoring, though more sophisticated implementations distinguish surface edits (typo corrections, formatting), substantive edits (reordering, factual corrections), and structural edits (paragraph-level rewrites) for richer signal.

The threshold below which the deployment has drifted toward tier 3 is organization-dependent and use-case-dependent. A reasonable starting point: if the median edit ratio across users falls below 30 percent, the deployment is functioning as model-produces-and-human-ratifies rather than as human-authors-with-AI-assistance. Below 15 percent, the tier-2 architectural claim is no longer supportable. These numbers are starting points; specific thresholds should be calibrated against the organization's working definition of *authored versus reviewed*.

The infrastructure requirement is the diff log itself. Most copilot-style products do not capture this by default. Microsoft Copilot in its current form, GitHub Copilot in default mode, and most sales-engagement copilots produce model output without storing the diff between produced output and sent version. The instrumentation gap is real and is the first thing to fix in a tier-2 deployment that does not have it.

A specific implementation note. The diff log contains user behavior data at fine granularity, which raises governance questions about retention and access. In employee-facing deployments operating in professional contexts, the legal basis for collection is generally clear under existing employment frameworks. In consumer-facing deployments, where the user is the recipient or a third party rather than the operator's employee, collection scope becomes a privacy question that belongs to legal counsel. The architecture requirement is diff logging; the scope of that logging is a governance decision.

Tier 3: reviewer latency per action

Tier 3's metric is the time between an action being presented to the human approver and the approval click. Implementation requires three log fields per action: the timestamp of presentation, the identity of the approver, and the timestamp of approval. The latency is the difference; the per-reviewer aggregation is the analysis.

Watch the distribution, not just the mean. A median of 8 seconds may be acceptable for a low-stakes outbound email; the same 8 seconds is mathematically incompatible with meaningful review of a customer credit decision. The threshold is action-type-specific and must be determined against the *minimum time a thoughtful reviewer would need* to read the action, consult its context, and form an independent judgment.

The threshold determination is itself worth discipline. Two methods work. The first is observational: watch reviewers process actions of each type and measure the time required for genuinely deliberate review; set the threshold at the lower bound of that distribution. The second is rational: estimate the cognitive components of meaningful review (reading the action, retrieving the context, evaluating against organizational criteria) and sum them. Both methods produce thresholds; using both and comparing surfaces calibration errors.

The compression signal is what matters. When the median latency falls *over time* — when a deployment that started with 12-second median latencies is at 4 seconds six months later — the deployment has drifted. Approval inflation has set in. The architectural claim of meaningful human involvement is no longer supportable.

The infrastructure requirement is the per-action presentation timestamp. Many approval-gate implementations log the approval event without the presentation event, which makes latency uncomputable. The instrumentation gap is the second-most-common gap I encounter in tier-3 audits, after the diff log gap at tier 2.

The regulatory implication, briefly here and at length in Module 6. Under the meaningful-human-involvement doctrine being articulated by EU data protection authorities under GDPR Article 22, an approval that the reviewer could not plausibly have considered is not a meaningful approval. The latency log is the evidence by which the organization demonstrates that approvals were meaningful. Without the log, the organization cannot prove its compliance posture; with the log, it can.

Tier 4: authoritative-record propagation latency

Before the metric itself, a connection worth making explicit. Module 1 named dashboard abandonment as the primary way a tier-4 deployment drifts toward tier 5 — the human reviewer stops watching, attention erodes, and the supervision the architecture assumed quietly stops happening. A reader expecting the tier-4 metric to measure dashboard attention directly will find this metric measures something else: how long an autonomous surface continues asserting a fact after that fact has changed. The two are tightly coupled. The entire purpose of the human watching the dashboard at tier 4 is to catch the system acting on stale information before the action propagates. When the dashboard is genuinely watched, stale assertions get caught and propagation latency stays low. When the dashboard is abandoned, propagation latency climbs, because nothing is catching the system asserting facts that are no longer true. Propagation latency does not measure the reviewer's attention; it measures the outcome that attention is supposed to produce. This is deliberate. Attention is difficult to measure and easy to fake — a reviewer can have the dashboard open and still see nothing. Propagation latency is computable from logs and tracks the failure mode that actually matters. It is the observable signature of tier-4 supervision, working or failed.

The dashboard does a second job that propagation latency does not capture. Beyond catching stale assertions, the tier-4 reviewer performs after-the-fact audit sampling — examining a sample of completed autonomous actions to judge whether the actions themselves were proper, not merely whether they were based on current data. A system can have flawless propagation latency and still take poorly judged actions: the customer record is current, and the outreach is still tone-deaf, mistimed, or aimed at the wrong contact. Action quality is the province of the evaluation harness described later in this chapter; the tier-4 reviewer's audit sampling is how completed autonomous actions enter that harness. Two things follow. First, propagation latency is the tier-integrity metric — the drift detector — but it is not a quality measure; action quality is measured by the harness, continuously, regardless of tier. Second, the audit sampling rate itself should be monitored. A reviewer who stops sampling is drifting the same way a reviewer who stops watching is drifting; declining sample coverage is dashboard abandonment in its second form, and it should trigger the same governance response.

Tier 4's metric is the dialer check from Module 1: when an authoritative record changes (a customer status, a price, a policy, an inventory level), how long until every autonomous surface that asserts about that record reflects the change. The 11x case where the AI phone dialer continued claiming ZoomInfo as a customer for weeks after the website was updated is the canonical instance of this metric failing.

Implementation is harder than the metric description suggests. It requires:

A registry of *authoritative records* — facts in source-of-truth systems that the agent reads from. Most organizations do not have this registry; building it is the first phase of tier-4 instrumentation. The registry should include: which fact, which system holds it as authoritative, which systems have downstream representations of it, and which agent surfaces assert about it.

A change-detection mechanism on each authoritative system. When the record changes, an event is emitted. This is standard CDC (change data capture) infrastructure for the source-of-truth side; the agentic dimension is connecting it to the agent surface inventory.

A propagation tracking mechanism. After the change event, monitor each downstream agent surface for when it reflects the new value. The gap between change-event timestamp and reflected-update timestamp is the propagation latency.

A threshold per record type. For high-stakes records (customer status, financial limits, regulatory determinations), the threshold should be measured in minutes. For lower-stakes records, hours may be acceptable. The threshold is a business decision informed by the blast radius of a wrong assertion at the time scale in question.

The organizational requirement here is more substantial than the technical one. Most enterprises have no central inventory of authoritative records, no formal mapping between source-of-truth systems and agent surfaces, and no ownership for keeping the mapping current as systems evolve. Tier-4 instrumentation surfaces this gap and forces its resolution. The resolution is the actual work; the metric is the instrument that compels it.

Tier 5: boundary-exception handling time

Tier 5's metric is the time between a boundary exception triggering and meaningful resolution. Implementation requires: the exception event log (when the agent hit the boundary, what action it was attempting, why the boundary engaged), the queue ingestion timestamp (when the exception entered the human handler queue), the assignment timestamp (when a specific human took ownership), and the resolution timestamp (when the exception was disposed — approved, rejected, modified, escalated).

Three computed metrics matter. First, queue dwell time — exception event to assignment. Second, assignment-to-resolution time — how long the assigned human took. Third, total handling time — exception event to resolution. The first surfaces capacity problems; the second surfaces individual handler problems; the third surfaces total system performance.

Watch in two directions. *Increasing* handling time signals queue overload — exceptions are accumulating faster than handlers can process them. The deployment is drifting toward tier 4: exceptions are stopping being handled in any meaningful sense, and the agent's behavior is effectively unbounded for the share of cases that reach the queue. *Decreasing* handling time, paradoxically, often signals exception rubber-stamping — handlers are clearing the queue by speed rather than by judgment. The deployment is drifting toward tier 6: the boundary architecture is becoming theatrical, and the agent is operating without meaningful constraint.

The threshold is bidirectional. A median handling time below a defined floor (such as 30 seconds for non-trivial exceptions) signals rubber-stamping. A 95th-percentile handling time above a defined ceiling (such as 4 hours for time-sensitive exceptions) signals overload. Both are tier-drift signals.

A specific implementation note. The exception log must capture not only what triggered the exception but what the handler decided and why. Without the why, exception patterns cannot inform boundary refinement. Without boundary refinement, the exception queue grows unbounded as the system encounters more situations the original boundary did not anticipate. The closing of the loop — exceptions inform refinements, refinements reduce the queue — is what makes tier 5 sustainable. Without the loop, tier 5 collapses into either tier 4 or tier 6 within months.

Tier 6: policy-drift detection latency

Tier 6's metric is the time between a policy update occurring and that update being evaluated against the system's intended behavior. Implementation requires the most demanding instrumentation of any tier.

The infrastructure: every change to the operational policy — whether through deliberate retraining, applied reward signals, in-context adaptation, or any other mechanism by which the agent's behavior changes — must produce a versioned policy artifact with a unique identifier and timestamp. The artifact must be retrievable; historical policy versions must be reproducible. This is the foundation; without it, nothing else in tier-6 evaluation is possible.

The detection mechanism: a behavioral evaluation suite runs against each new policy version, comparing outputs on a fixed set of probes against the prior version's outputs and against the original deployment baseline. Drift is the signed magnitude of the change. Rapid drift on safety-relevant probes triggers alert; sustained drift triggers review.

The probe design is non-trivial. The Anthropic November 2025 paper on emergent misalignment from production reward hacking established that drift on a specific reward signal can generalize to behaviors the operator did not anticipate — alignment faking, malicious-actor cooperation, sabotage attempts. The probe set must include not only the obvious dimensions (accuracy, cost, latency on representative tasks) but adversarial probes designed to surface the generalization patterns the Anthropic research documented. Most enterprise organizations cannot build this probe set themselves; the realistic options are external evaluation services, open-source probe libraries (which are emerging but immature), or partnership with the model provider.

The detection latency is the gap between policy update and probe evaluation. A bound on this latency is what makes tier 6 operationally governable. An unbound detection latency means the system can operate in a drifted state for an arbitrary period before the drift is observed; for most enterprise use cases, an arbitrary period of misaligned operation is not an acceptable risk posture.

Operating tier 6 without this infrastructure is operating without an instrument. The deployment may be producing acceptable outputs today; the operator has no basis to assert it will produce acceptable outputs tomorrow.

Section 2: The integrity dashboard

Each tier-integrity metric is a number. Each number lives in a dashboard. The dashboard is the instrument by which the cadence governance from Module 1 actually catches drift; without the dashboard, the cadence is theatrical.

Three principles for the integrity dashboard.

First, the dashboard shows distributions over time, not point values. A median reviewer latency of 7 seconds today is meaningful only against the trend over the past 30 days, 90 days, six months. The dashboard's primary visual unit is the time series, not the gauge.

Second, the dashboard surfaces alerts at thresholds, not just data. The threshold for each tier-integrity metric is set when the deployment is documented (the threshold appears in the architecture decision record), and the dashboard fires alerts when the metric crosses the threshold. The alert routes to the named owner of the deployment's ADR; the alert's existence is the cadence governance's connection to action.

Third, the dashboard reports the *tier diagnostic* alongside the metrics. The diagnostic is the answer to *given these metrics, what tier is this deployment actually operating at?* The diagnostic resolves the seven-question diagnostic from Module 1 against current data: if reviewer latency has compressed below threshold, the diagnostic resolves tier-3 deployments to operating-at-tier-4; if the boundary-exception handling time is below the rubber-stamping floor, the diagnostic resolves tier-5 deployments to operating-at-tier-6. The diagnostic is what makes the dashboard actionable for a CTO who is not in the metrics every week — the diagnostic translates the metrics into a tier label that connects directly to the governance posture.

A specific implementation note. The dashboard should be visible to the named owner, the operational team running the deployment, and the governance function (typically risk, legal, or a CAIO function where one exists). Restricting the dashboard to engineering reduces its governance value to zero; making it visible only to executives reduces its operational value to zero. The dashboard is shared infrastructure across the cadence levels.

Section 3: Drift detection beyond the metrics

Tier-integrity metrics are the load-bearing diagnostic. They are not exhaustive. Two additional drift-detection mechanisms warrant attention.

Behavioral red-teaming. Periodically — quarterly is reasonable for tier 4 and 5 deployments, monthly for tier 6 — a small team probes the deployment with adversarial inputs designed to surface drift the metrics might miss. The probes should include: edge cases the original system handled correctly (re-test that they still do), known failure patterns from public incidents (re-test that the system is not vulnerable), and novel inputs designed to elicit the generalization patterns from the Anthropic research at tier 6. The output is a report on the system's current robustness, with deltas from prior reports as the drift signal.

Sampled output review. A small percentage of production outputs — typically 1 to 5 percent depending on volume — is reviewed by a human evaluator with domain expertise, against a defined rubric. The review captures whether the output was correct, whether it would have been an acceptable output six months ago, and whether the agent's reasoning (where exposed) reflects the patterns the deployment was designed to encode. The sampled review surfaces drift in dimensions the metrics cannot — particularly tone, nuance, and judgment quality — and provides the qualitative complement to the quantitative integrity dashboard. The sampling rate itself is a monitored quantity. The review depends on a defined percentage of outputs actually being sampled; when that coverage declines — because the reviewing capacity is absorbed by other work, because volume grew and the sample count did not — the drift-detection mechanism is quietly failing even though no individual review went wrong. Declining sample coverage is the second form of dashboard abandonment named in Section 1: the tier-4 reviewer who stops sampling completed actions is drifting the same way the reviewer who stops watching live is drifting. Coverage is tracked alongside the review findings, and a sustained drop triggers the same governance response as a threshold crossing on any tier-integrity metric.

Both mechanisms cost real money and engineering time. Both are necessary at higher tiers. The honest position: an organization that is not willing to invest in red-teaming and sampled review is not willing to invest in the governance overhead that operating at tier 4 or above actually requires. The architectural choice and the evaluation investment are the same choice.

Section 4: The evaluation cadence

The cadence-governance frame from Module 1 has direct implications for how evaluation is organized. Four levels, each with specific evaluation activity.

Weekly: integrity dashboard review. The team operating the deployment reviews the tier-integrity metric distributions over the past week. The review takes 15 minutes. It asks one question: *is the trajectory consistent with the tier we committed to operating at?* When the answer is yes, the review ends. When no, the issue moves to the monthly review. The weekly review's primary function is catching threshold crossings as they happen rather than discovering them in retrospect.

Monthly: trend and exception review. The named owner of the deployment's ADR examines the four-week trend on the integrity metrics, the exception patterns from the operational logs, and any incidents from the period. The review takes 30 minutes. It produces a specific action: adjust the threshold, expand the reviewer capacity, escalate to quarterly. The monthly review's primary function is connecting the weekly observations to operational change.

Quarterly: red-team review. The behavioral red-team conducts its quarterly probe; the sampled output review is reported; the integrity dashboard's longer-term trends are examined. The review takes 60 minutes. It includes the CTO, the CPO, the operational owner, and the function responsible for risk. It produces decisions about deployment continuation, tier-claim adjustment, vendor performance, and regulatory posture. The quarterly review is the level at which tier-claim decisions are revisited.

Semi-annual: strategic review. The board or senior governance forum examines the integrity dashboards across the deployment portfolio, the cumulative regulatory exposure, and the proportionality of evaluation investment to deployment scale. The review takes 90 minutes, twice per year. It addresses portfolio-level questions: are we operating at the right tiers, are our evaluations adequate to the stakes, are we investing proportionately. The semi-annual review's function is keeping evaluation investment calibrated to the organization's actual risk posture.

The cadence is not theater if it is connected to action. Each level has triggers that escalate when conditions cross thresholds. Without the triggers, the cadence is performance; with them, the cadence is the mechanism by which evaluation translates into governance.

Section 5: Performance evaluation across the tiers

Tier-integrity is the layer this chapter contributes. Performance evaluation is the layer the existing literature handles well, and the chapter's posture is to show how the existing frameworks apply differently across tiers rather than to replace them.

The four existing frameworks worth referencing.

CLEAR (Cost, Latency, Efficacy, Assurance, Reliability). Published November 2025. The strongest contribution is on cost — the framework documents that benchmarks systematically under-report cost, with leading agents showing 50x cost variation for similar accuracy. The cost dimension is load-bearing for tier 4 and above, where high-volume autonomous execution makes cost a primary operational concern.

Aisera CLASSic (Cost, Latency, Accuracy, Stability, Security). Demonstrates that domain-specific agents reach 82.7 percent accuracy versus 59-63 percent for general LLMs at 4.4-10.8x lower cost. Useful at tier 3 and above where deployment economics matter; particularly useful for vendor evaluation, treated in Module 4.

MOYA four-pillar (LLM, Memory, Tools, Environment). Distinguishes runtime behaviors that static testing misses — *"deviations from expected policies or validation flows remained undetected by existing assessment methods as they surfaced only during runtime execution rather than in static validation phases."* The runtime focus is essential at tier 4 and above where the agent is operating in the world and static validation cannot capture the production environment.

QuantumBlack three-layer (model, single agent, multi-agent system). Distinguishes evaluation across the SDLC phases (planning, build, operation). Useful for organizing the eval harness as a phased instrument rather than a single tool — different evaluations run at different points in the development and operation lifecycle.

These frameworks are complementary, not competing. A mature evaluation harness draws from all four. The contribution of this chapter is the layer above them: the tier determines *which dimensions of which framework matter most* and *what the evaluation cadence and depth should be* for the specific deployment.

The tier-specific application of performance evaluation:

At tier 1 and 2, performance evaluation focuses on output quality — does the model produce useful suggestions, does the edited artifact meet the organization's quality bar. The CLEAR/CLASSic accuracy and quality dimensions apply directly. Cost and latency are secondary because volumes are typically low and the human is in the artifact production loop.

At tier 3, performance evaluation expands to include reviewer effectiveness — does the approval step catch problems the model produced. The evaluation must capture both model performance and reviewer performance separately, because either can be the source of failures. The MOYA environment pillar becomes relevant: how does the system behave in the production review interface as opposed to the test interface.

At tier 4, performance evaluation expands again to include downstream effect tracking. Because the system is acting in the world, the evaluation must capture not only the action but its consequences. Did the email get a response? Did the database write produce the expected downstream state? Did the customer service interaction resolve the underlying issue? Outcome metrics are first-class at this tier, alongside output metrics. The cost dimension from CLEAR becomes load-bearing because volumes are typically high.

At tier 5, performance evaluation includes boundary effectiveness. Was the boundary architecture catching the cases it was designed to catch? Were exceptions being escalated to handlers with the information they needed? The MOYA environment pillar and the QuantumBlack multi-agent layer apply directly when the bounded autonomous system coordinates with other agents or tools.

At tier 6, performance evaluation must include drift effects on system behavior. The Anthropic research established that policy adaptation can produce behaviors that were not present at deployment. Performance evaluation at tier 6 is not a snapshot; it is a continuous comparison of current behavior against deployment baseline, with attention to the generalization patterns documented in the research.

The point is not that the existing frameworks are wrong. The point is that the dimensions of evaluation that matter most shift across tiers, and an evaluation harness that applies all dimensions equally at all tiers either over-instruments low tiers or under-instruments high tiers. The tier determines the proportionality.

Section 6: The 37 percent gap, addressed

The lab-to-production gap from earlier in the chapter has specific implications for how the evaluation harness is structured.

The harness must operate continuously in production, not only as a pre-deployment gate. Pre-deployment evaluation is necessary — a system that fails benchmark testing should not reach production — but the gap means pre-deployment evaluation is insufficient as a guarantee of production behavior. The instrumentation has to follow the system into production and continue measuring there.

Continuous production evaluation has a structure that pre-deployment evaluation does not. It must work against actual workloads, not curated benchmarks. It must capture user behavior, not just system behavior. It must accumulate evidence over time rather than snapshot performance. The MOYA framework's runtime focus is the closest of the existing frameworks to this requirement; the chapter's tier-integrity layer is the missing piece.

The honest implementation reality. Continuous production evaluation is more expensive than pre-deployment evaluation. It requires infrastructure that runs alongside the system in production, with overhead that the production environment must absorb. Some of the overhead is irreducible: the diff log at tier 2, the propagation tracking at tier 4, the policy versioning at tier 6 are all instrumentation costs that the organization must pay if it wants to operate at the corresponding tiers.

The cost-versus-benefit framing matters. The cost of evaluation infrastructure is real. The cost of operating without it is also real, and shows up as: regulatory liability when the meaningful-human-involvement claim cannot be supported, reputational damage when the dialer-style failure surfaces, contractual exposure when the vendor representation cannot be verified. An organization that resists evaluation infrastructure investment is choosing to absorb the latter costs implicitly. The choice is whether to pay measurably or unmeasurably.

Section 7: Vendor-provided evaluation versus operator-built evaluation

A specific question that arises in every tier-3 and above deployment: should the operator build their own evaluation infrastructure, or rely on the vendor's?

Vendors increasingly offer evaluation tooling alongside their agentic AI products. Galileo positions itself in this space, as do most major vendors of agent platforms (LangChain via LangSmith, AutoGen via AutoGen Bench, others). The vendor-provided evaluation has real advantages: it is integrated, it captures dimensions the operator would not easily capture, and it is typically less expensive than building equivalent infrastructure independently.

The vendor-provided evaluation also has structural limitations that an operator depending on it should understand.

The vendor evaluates against the dimensions the vendor cares about. A vendor's evaluation harness optimizes for what the vendor wants to demonstrate, which is typically performance against benchmarks that show the vendor's product favorably. The operator's evaluation should optimize for what the operator's regulators, customers, and board care about, which is not always the same set of dimensions.

The vendor's evaluation does not capture the operator's tier-integrity. The vendor cannot know whether the operator's reviewers are rubber-stamping, whether the operator's authoritative records are propagating, whether the operator's exception queue is being meaningfully handled. These are operator-side measurements that no vendor can produce because they involve operator-side behavior.

The vendor's evaluation cannot serve the operator's regulatory defense. Under GDPR Article 22 and the EU AI Act, the operator (deployer in AI Act terms) is the entity responsible for demonstrating compliance with meaningful-human-involvement requirements. The evidence base for that defense must be the operator's evidence, captured by operator-controlled instrumentation. Vendor-provided evidence is supplementary; it cannot be the primary record.

The recommendation: use vendor evaluation infrastructure for performance dimensions where the vendor has natural visibility (model behavior, tool invocations, retrieval quality) and build operator-controlled infrastructure for tier-integrity dimensions and regulatory-defense dimensions. The split is not 50/50; it depends on the tier and the specific deployment. But the organization that depends entirely on vendor evaluation is depending on an instrument that does not capture what the organization actually needs to know.

Section 8: The evaluation harness as integrated instrument

The chapter has treated tier-integrity evaluation and performance evaluation as separate threads. The eval harness as an integrated instrument combines them.

The integrated harness has four components.

The integrity layer captures the tier-integrity metrics, runs the dashboard, fires the threshold alerts, and produces the tier diagnostic. This is the layer this chapter contributes to the literature.

The performance layer captures the dimensions from CLEAR, CLASSic, MOYA, or whichever framework the operator has chosen, scaled to the tier of the specific deployment. This is the layer the existing literature handles well.

The drift layer runs the red-team probes, the sampled output reviews, and the policy-version evaluations at tier 6. This is the layer that catches drift the metrics miss.

The cadence layer routes the outputs of the other three layers into the weekly/monthly/quarterly/semi-annual review structure. This is the layer that connects evaluation to governance.

The four layers form a single instrument. They share the underlying logging infrastructure. They share the dashboard. They share the alert routing. They share the named owner. The harness is not four separate tools; it is one instrument with four operational functions.

A specific implementation guide for new deployments. Build the layers in order. The integrity layer first, because it is the simplest to instrument and produces immediate value. The performance layer second, scaled to the tier and use case. The drift layer third, when the deployment has reached the maturity to support red-teaming and sampled review. The cadence layer fourth, connecting the prior three into governance.

For deployments already in production without an eval harness, build in reverse priority. The cadence first, because cadence is what catches issues even with imperfect data. The integrity layer second, because it is the load-bearing diagnostic. The performance layer third, leveraging vendor tools where available. The drift layer fourth, when the operational maturity supports it.

Either build order works. The wrong choice is to wait until you have time to build all four; that time does not arrive, and the deployment operates without the harness in the meantime.

Monday artifacts

CTO artifact: Eval harness architecture decision record

A one-page document that names the four layers of the harness, the named owner of each layer, the integration points between them, and the specific tier-integrity metrics being captured for the deployment. The ADR's primary function is to make explicit what is being measured and what is not, so that gaps in the evaluation surface as architectural decisions rather than as discovered failures.

The required sections. *What this harness measures* — the list of tier-integrity metrics, performance dimensions, drift probes, and cadence levels covered. *What this harness does not measure* — explicit gaps, with the reason and the planned remediation date. *Owners* — the named human responsible for each layer, with the named human responsible for the harness as a whole. *Integration points* — how each layer connects to the others, where the data flows, where the alerts route. *Review cadence* — when each layer's outputs are reviewed at each cadence level. *Cost and resourcing* — what the harness costs to operate, both in vendor spend and in human time, and what the budget commitment is.

The ADR is reviewed quarterly. The harness evolves; the ADR captures the evolution.

Architect artifact: Logging schema specification

The technical specification for the logs that the harness depends on. The specification must include, at minimum:

For tier-integrity logging: the diff log structure at tier 2, the action presentation and approval log structure at tier 3, the authoritative-record-and-propagation log structure at tier 4, the exception event and disposition log structure at tier 5, the policy-version log structure at tier 6.

For performance logging: the action ID and trace structure, the cost capture per action, the latency capture per action, the outcome tracking where applicable.

For drift detection: the probe execution log, the sampled output review log, the policy-version evaluation log.

The specification should be implementation-ready — schema definitions, retention policies, access controls, query patterns. The architect should be able to hand it to an engineering team and have the logs flowing within sprints.

The retention question deserves specific treatment. The logs are evidence. The retention period should be the longest of: the relevant statute of limitations for the workloads the system handles, the regulatory retention requirements (which for high-risk EU AI Act systems may be substantial), the organization's own policy needs. Most organizations default to short retention periods designed for cost minimization; for evaluation logs, this is the wrong default. Plan for years, not months.

CPO artifact: Evaluation investment proportionality

The CPO's specific concern with evaluation is proportionality — is the organization investing in evaluation at a level commensurate with the risk posture of its deployments. The artifact for this is a portfolio-level review, conducted semi-annually, that asks four questions.

What deployments do we operate, at what tiers? The tier inventory across the AI portfolio.

What evaluation infrastructure does each deployment have? The harness inventory, with gaps named explicitly.

Does the evaluation investment match the tier? A tier-5 deployment with the evaluation infrastructure of a tier-3 deployment is under-invested; the under-investment is a governance liability the CPO owns.

Where are we under-investing, and what is the remediation plan? The output of the review is a portfolio-level remediation roadmap with specific commitments and dates.

The proportionality question is a CPO concern because it cuts across deployments and connects to the organization's overall risk posture. Engineering teams optimize their individual deployments; legal optimizes against specific regulations; risk optimizes against the framework du jour. None of these naturally produces the portfolio view. The CPO does, or it does not happen.

Forward references

Module 3 (Logging, Reversibility, and Incident Response) covers the logging schemas at deeper implementation depth, including the session-level reconstruction patterns, and the runbook integration with the integrity dashboard. The schemas this chapter named are specified in detail in Module 3.

Module 4 (Procurement and Vendor Evaluation) covers the vendor-provided-versus-operator-built question at deeper depth, including specific contract language for evaluation provisions, audit rights, and tier-claim verification. Vendors who refuse audit rights against operator-controlled evaluation infrastructure are signaling a position that has procurement implications.

Module 5 (Organizational Design) covers the named ownership of the eval harness across the cadence levels, including the question of whether the harness ownership belongs to the CTO, the CAIO where one exists, the operational owner, or the risk function. The chapter takes a position that may be controversial.

Module 6 (Regulatory Posture) covers the regulatory implications of evaluation infrastructure in detail, particularly the GDPR Article 22 evidence requirements, the EU AI Act technical documentation requirements (Article 11), and the conformity assessment process for high-risk AI systems. The integrity dashboard and the performance harness are load-bearing artifacts in the regulatory defense; Module 6 specifies what makes them defensible.

Module 3: Logging, Reversibility, and Incident Response

The how. What you log so the system is auditable, what reversibility primitives you build so the system is recoverable, and what runbooks you write so the system is survivable when it fails.

The recommendation, stated first

Build the logging infrastructure before you need it. Build the reversibility primitives before you need them. Write the incident runbooks before you need them. None of these are optional for agentic AI deployments at tier 3 or above, and all three are the kind of infrastructure that, once needed, cannot be retrofitted in time.

The honest framing question for this chapter comes from a piece on enterprise agentic AI incident response that surfaced in early 2026: *who owns the agent at 2am?* The standard SRE runbook assumes a microservice with an owning team, a deployment pipeline, an on-call rotation. Agentic AI deployments often have none of these in clear form. The agent was built by a project team that has moved on, sits inside a vendor product whose support boundary excludes operator-side configuration, draws context from systems owned by different functions, and produces effects that touch teams who never knew the agent existed. When the agent does something wrong at 2am, the question of *who is responsible for fixing it* has no obvious answer, and the time spent finding the answer is time the incident is compounding.

This chapter's premise is that the answer must be established *before* the 2am call. The infrastructure, the architecture, and the runbook all converge on the same requirement: clarity about ownership, observability, and recovery, designed in advance and tested in advance, so that the 2am response is execution rather than discovery.

Section 1: Logging — the infrastructure

The chapter's first section is the foundation everything else depends on. Without the logs, reversibility is impossible and incident response is guesswork.

What's distinctive about agentic AI logging

Standard application logging captures discrete events: a request came in, a service responded, a status code was returned. Agentic AI logging has to capture something different: the *reasoning trace* of a system that took multiple steps, made multiple tool calls, retrieved multiple pieces of context, and produced a final output through a process that may not be reproducible.

The implication is that agentic AI logs are larger and more structured than standard application logs. A single user-facing interaction can produce dozens of internal log entries — model calls, tool invocations, retrieval queries, reasoning steps, intermediate outputs, downstream effects. The log structure has to support session-level reconstruction, which is the property that distinguishes useful agentic AI logs from logs that produce more noise than signal.

Three properties matter for any logging schema at any tier.

Session correlation. Every event produced by a single user interaction must be linkable to the parent session. The session ID is the primary key by which incident response reconstructs what happened. Without consistent session correlation across the model, the orchestration layer, the tool integrations, and the downstream effects, the logs cannot be assembled into a coherent narrative when needed. OpenTelemetry conventions provide the technical scaffolding; the discipline is ensuring every component participates in the same trace.

Causal ordering. Within a session, the order of events matters. The logs must support reconstructing not just what happened but in what order, including parallel branches when the agent ran tools concurrently. Sub-second timestamp precision is necessary, not optional. Logical-clock approaches (Lamport timestamps, vector clocks) become relevant when the agent coordinates across distributed components.

Sufficient context. Each log entry must contain enough context to be interpretable in isolation if needed. A tool invocation log that records *tool X called with parameters Y* is useful; one that records *a tool was called* is not. The context discipline is to capture the inputs that determined the agent's choice, not just the choice itself.

Tier-specific logging schemas

The properties above apply at every tier. The specific fields and the retention policies vary.

At tier 1 (suggestion). Minimal logging requirement. The model produces output into a separate surface; the user composes from scratch or copies deliberately. The relevant log is the suggestion-surface usage log: what suggestions were produced, when they were displayed, and — where the architecture supports it — whether they were copied into the artifact. Most products do not log the copy event because the copy is a normal text-input action that the product cannot distinguish from typing. The honest acknowledgment: tier 1 is not architecturally instrumentable for tier-integrity purposes, and the logging serves model-improvement and product-analytics goals rather than governance goals.

At tier 2 (draft-and-edit). The diff log is the core artifact. Schema:

```
session_id: string (UUID)
```

```
user_id: string (or anonymized identifier per privacy
posture)

artifact_type: enum (email, code_file, document, message)

model_version: string

prompt_template_version: string

model_output: text (the content the model produced into
the composition surface)

sent_version: text (the content the user sent or
committed)

edit_events: array of {timestamp, position, change_type,
before, after}

sent_timestamp: timestamp

metadata: object (channel, recipient context, business
workflow)
```

The retention period for the diff log should be the longer of the relevant statute of limitations and any sector-specific regulatory requirement. For most commercial communications, this is several years. The diff log is evidence, not just operational data.

A specific implementation note. The diff log captures user behavior at fine granularity, which raises governance questions about access. Engineers debugging a model issue do not need access to specific user diffs; legal counsel evaluating a discrimination claim might. The access model should be tiered, with broad aggregate access for engineering and narrow individual-record access for legal and compliance functions on a documented-need basis.

At tier 3 (approve-and-send). The approval log adds three fields to the tier-2 baseline:

```
approver_id: string

presentation_timestamp: timestamp

approval_timestamp: timestamp
```

```
approval_decision: enum (approved, rejected, escalated)

review_interface_state: object (what the approver could
see at decision time)
```

The presentation timestamp is the most-commonly-omitted field and the most important one. Without it, reviewer latency is uncomputable; with it, the rubber-stamping pattern becomes detectable. The review-interface-state field is the second-most-important: it captures what information the approver had access to when making the decision, which becomes evidence in any subsequent regulatory inquiry about whether the approval was meaningful.

At tier 4 (supervised execution). The full decision-trace log structure:

```
session_id: string

action_id: string (globally unique)

parent_action_id: string (when actions chain)

agent_version: string

policy_version: string

inputs: {prompt, retrieved_context, system_state,
prior_session_state}

reasoning_trace: array of reasoning steps (where the model
exposes them)

tool_invocations: array of {tool, parameters, result,
timestamp, latency, cost}

final_output: object (the action taken)

downstream_effects: array of {target_system, change,
change_id, timestamp}

authoritative_record_reads: array of {record_type,
record_id, value_read, source_system, source_timestamp}

authoritative_record_writes: array of {record_type,
record_id, value_written, target_system}
```

The authoritative-record fields are what enable the dialer check. Every assertion the agent makes about a fact in the world must be linkable to the source-of-truth read that produced it. When the source-of-truth changes, every downstream agent surface that has previously asserted from the old value must be findable through the log; this is what makes the dialer-style failure detectable rather than discoverable-by-accident.

The retention requirement at tier 4 is more substantial than at lower tiers. The logs are the regulatory defense for the deployment's compliance posture; they are also the evidence base for any incident reconstruction. Plan retention in years, with longer windows for high-stakes use cases.

At tier 5 (bounded autonomy). The tier-4 schema plus boundary-evaluation logging:

```
boundary_evaluation_id: string

boundary_version: string

considered_action: object

permitted: boolean

reasoning: text (why the boundary engaged or did not
engage)

exception_raised: boolean

exception_id: string (when applicable)
```

Every action the agent considered — not just the actions it took — must be linkable to a boundary evaluation. This is the field most commonly missing in tier-5 deployments, and its absence makes boundary-effectiveness evaluation impossible. An agent that hits a boundary 1,000 times and escalates 50 of them is qualitatively different from an agent that hits the boundary 50 times; both produce 50 escalation events, but only the boundary-evaluation log distinguishes them.

The boundary-version field allows historical reconstruction: when an exception fires today, what version of the boundary engaged? When a new boundary version is deployed, what does behavior look like immediately afterward? Boundary changes are operational events with consequences, and the logs must capture them as such.

At tier 6 (self-modifying autonomy). The tier-5 schema plus policy-versioning infrastructure:

```
policy_version: string (every distinct policy state)

policy_lineage: array of (parent_policy_version,
derivation_method, derivation_timestamp)

training_data_summary: object (what data informed this
policy version)

evaluation_results: array of (probe_set, baseline,
current, delta)

deployment_status: enum (active, retired, quarantined)
```

The policy-lineage tracking is what makes drift characterization possible. When the current policy is six versions removed from the deployment baseline, the trajectory through those six versions matters; understanding which adaptations cumulated into the current behavior is the substance of drift analysis at tier 6.

Implementation honesty: this infrastructure is not available in most enterprise deployments. The organizations operating it are primarily AI research labs and a small number of vendors who built it specifically because they understood the tier they were operating at. For an enterprise considering tier-6 operation, the question is whether to build this infrastructure (substantial undertaking), partner with a vendor that has it (acceptable for some use cases, not for others), or operate at tier 5 instead. The chapter's recommendation throughout has been the third option for most enterprise deployments.

Logging as evidence

A specific framing worth being explicit about. The logs at tier 3 and above are not primarily operational data. They are evidence — for regulatory defense, for incident reconstruction, for contractual disputes with vendors, for internal post-mortems that drive system improvement. Treating them as evidence has implementation consequences.

Evidence must be tamper-evident. Append-only log structures, cryptographic hashing, or write-once storage prevent retroactive modification. Most organizations log to systems where retroactive modification is technically possible; for evidence purposes, this is a defect. The remediation is access control plus audit logging on the audit logs themselves.

Evidence must be retrievable on demand under time pressure. A log archive that takes three weeks to query is operationally adequate for trend analysis and useless for incident response. Tiered retention — recent logs hot, older logs warm, archived logs cold but retrievable within defined SLAs — is the standard approach. The SLA targets should match the use cases: minutes for incident response, hours for regulatory inquiry, days for litigation.

Evidence must be reproducible against the system that produced it. A log that captures *the agent did X based on input Y* is evidence only if the agent can be reconstructed in the state it was in when the log was written. The agent-version, policy-version, and prompt-template-version fields in the schemas above serve this purpose. Without versioning, the logs become uncorroborated assertions about a system that no longer exists.

Section 2: Reversibility — the architectural property

Logging captures what happened. Reversibility determines what can be undone. The two are connected: reversibility primitives that operate without supporting log infrastructure are ineffective, and log infrastructure without reversibility primitives is documentation of damage rather than recovery.

What reversibility means at each tier

At tier 1 and 2, reversibility is approximately standard communication-retraction. The artifact has been sent; if it was wrong, the same mechanisms apply that would apply to any wrong human-authored artifact. The agentic dimension adds nothing distinctive.

At tier 3, reversibility remains per-artifact for the individual approval. The agentic dimension introduces the pattern-level question: when the approval gate has been operating ineffectively for weeks (the rubber-stamp problem), reversing the pattern requires first detecting it. The reversibility primitive is the integrity dashboard from Module 2 — without continuous monitoring, the pattern reverses no faster than the next external trigger forces examination.

At tier 4, reversibility becomes a real engineering problem. Three classes of action require three different reversibility primitives.

Communications. Email retraction protocols, message retraction APIs where supported, correction-and-apology workflows where retraction is impossible. Most modern messaging systems support some form of retraction or unsend within bounded time windows; the architecture should leverage these where available and provide compensating mechanisms where not. The cost of retraction is reputational — sending an apology email is itself a customer-facing event — and the architecture should treat retraction as an action in its own right, with its own logs and approvals.

Database writes. Transactional boundaries that include the agent's actions as first-class operations. Compensating writes when full transactional rollback is impossible. Idempotency keys that support retry without duplication. The agent's writes should be designed with the same rigor as any production code's writes; the fact that the writer is an agent does not relax the requirement.

Third-party API calls. Compensating actions defined for every integration. When the agent calls an external API, the architecture must know how to call the corresponding undo (refund, cancel, deactivate) when needed. Some APIs do not have compensating operations; for those integrations, the agent's actions are not reversible in the engineering sense, and the architectural choice to use them at tier 4 must be made deliberately rather than discovered after the fact.

The dialer check from Module 1 is the specific tier-4 reversibility primitive worth implementing first. Every authoritative record the agent reads from must have a defined propagation path to every agent surface that asserts about it, with bounded latency from authoritative-record change to surface update. The propagation infrastructure is the technical implementation of the dialer check; the metric from Module 2 is the monitoring of it.

At tier 5, reversibility splits between action rollback (inheriting from tier 4) and *boundary rollback*. Boundary changes are operational events. When a new boundary version is deployed and produces unintended behavior, the rollback is reverting to the prior boundary version, with consideration for actions taken in the interim. Implementation requires:

- Boundary versions stored with full provenance.
- Deployment mechanisms that support rollback to specific historical versions.
- Monitoring of boundary-change effects with alerting when post-change behavior diverges from expected.
- Defined ownership of rollback decisions, with named humans authorized to trigger them.

The boundary-rollback primitive is the tier-5 equivalent of database rollback at tier 4. Most organizations treat boundaries as configuration rather than as code, which means they lack the rollback infrastructure that production code requires. Treating boundaries as code is the architectural shift that makes tier-5 governance operational.

At tier 6, reversibility splits further into *action rollback*, *policy rollback*, and *downstream-effect remediation*. The first two are technical operations that the policy-versioning infrastructure supports. The third is operational archaeology: when the agent has operated under a drifted policy for weeks, the actions taken during that period reflect the drift in ways that may not be individually large but are collectively significant. Cleaning up the cumulative effect is not a technical operation; it is an organizational one, requiring identification of affected interactions, customer-by-customer remediation, and potentially regulatory disclosure.

The honest position: tier-6 reversibility is the hardest problem in this chapter, and the literature does not yet have a clean solution. The Anthropic research from late 2025 demonstrates that drift can produce behaviors the operator did not anticipate; the operational question of how to remediate the cumulative effects is downstream of that demonstration and is being actively figured out. Operators at tier 6 should plan for reversibility costs that exceed the lower tiers by an order of magnitude, and should treat the planning itself as part of the architectural decision.

The reversibility audit

A specific exercise worth running at every tier-4-and-above deployment: the reversibility audit.

The audit asks, for each action class the agent can take: *what is the retraction primitive, what is its detection-to-rollback latency target, who is authorized to trigger it, when was it last tested?* The output is a matrix that surfaces gaps where retraction primitives do not exist, where latency targets have not been defined, where ownership is ambiguous, where testing has been skipped.

The audit is not theoretical. The matrix produces actions: build the missing primitives, define the missing targets, assign the missing owners, run the missing tests. An organization that has run the audit and remediated the gaps is operationally prepared for tier-4 incidents. An organization that has not run the audit is hoping it will not need the primitives, which is a posture that holds until the first incident proves it does not.

The audit should run quarterly at tier 4 and 5, monthly at tier 6. The auditing function typically belongs to the same group that owns the integrity dashboard from Module 2, because the two are coupled: the dashboard surfaces issues that need reversal, and the audit ensures the reversal infrastructure exists.

Section 3: Incident response — the operational practice

Logging and reversibility are infrastructure. Incident response is the operational practice that draws on both when something goes wrong. This section gives the runbook structure.

What's distinctive about agentic AI incidents

Standard incident response assumes incidents have discrete causes that can be isolated to a specific failure: a service crashed, a deployment failed, a database query timed out. The runbook is structured around the principle of localizing the cause and reverting or fixing the specific failure point.

Agentic AI incidents often do not have discrete causes. The agent's behavior emerges from the interaction of the model, the prompts, the retrieved context, the tool integrations, and the user's input pattern. A single bad output may be the result of multiple contributing factors, each individually within normal range but collectively producing the failure. The standard runbook structure breaks against this — there is no single failure to localize.

Three implications for runbook design.

Session-level reconstruction is the first response action, not the third. When standard service incidents fire, the on-call engineer typically checks dashboards, examines recent deployments, and looks for anomalies. When agentic AI incidents fire, the on-call engineer must first reconstruct the specific sessions in which the failure manifested — the full trace of what the agent saw, what it considered, what it produced. Without the reconstruction, the engineer cannot determine whether the failure was a model issue, a context issue, a tool integration issue, or an input distribution shift. The session-level logs from Section 1 are what makes this reconstruction possible.

Severity classification is non-standard. Standard service incidents are classified by impact (number of users affected, revenue loss per minute, customer-facing visibility). Agentic AI incidents add a dimension: *was the agent's behavior* *outside its declared tier?* An agent that produced a wrong but normal-tier output is a different incident from an agent that produced a tier-violating output. The severity classification must capture both the impact dimension and the tier-violation dimension, because they trigger different response paths.

Containment options are tier-specific. Standard service incidents can be contained by traffic shifting, feature flags, or service degradation. Agentic AI incidents have additional containment levers: dropping the agent to a lower tier (engaging a more conservative approval gate, tightening boundaries, suspending self-modification), narrowing the input distribution (restricting to a subset of users or use cases), or triggering full circuit-breaker shutdown. The runbook must specify which containment lever is appropriate for which incident type, because picking the wrong lever can either over-respond to a minor incident or under-respond to a serious one.

The runbook structure

A runbook for agentic AI incidents has eight sections, each with tier-specific content.

Detection criteria. What signals indicate an incident is in progress? At tier 4, this includes integrity-dashboard alerts, customer complaint patterns, downstream system anomalies, and red-team probe results. At tier 6, it adds policy-drift signals, evaluation-suite regression, and the specific generalization patterns from the Anthropic research. The detection criteria must be specific enough that the on-call engineer can determine *whether this is an incident* without prolonged analysis.

Severity classification. Two-dimensional, as noted above. Severity-by-impact is standard; severity-by-tier-violation is new. The runbook should provide a matrix that combines them and routes to specific response paths.

Initial assessment actions. What does the on-call engineer do first? For agentic AI incidents, the first action is almost always session-level log retrieval and reconstruction. The runbook should specify the queries, the dashboards, the tools, and the time bounds for the initial assessment.

Containment options. The tier-specific containment levers, with criteria for choosing among them. The runbook should be explicit that containment may include dropping to a lower tier; this is the agentic-AI-specific containment that standard SRE runbooks do not have.

Stakeholder notification. Who needs to know, when, and at what level of detail. Agentic AI incidents often touch teams (legal, compliance, customer success, vendor relations) that standard service incidents do not. The notification list at tier 4 and above includes these teams as standard, not as escalation.

Investigation framework. A structured approach to determining root cause. Hamming's 4-Stack approach to voice agents — telephony, audio, intelligence, output, checked bottom-up — is one example of a tier-specific framework. The runbook should provide an analogous framework for the deployment in question, recognizing that 60 percent of voice-agent failures are in the lower stacks (per Hamming's data) and that similar concentration patterns exist for other agent types. The investigation framework prevents the on-call engineer from jumping to model debugging when the actual failure is in tool integration or context retrieval.

Recovery actions. The reversibility primitives from Section 2, applied to the specific incident. Communications retraction, database rollback, third-party compensation, boundary rollback, policy rollback — the runbook specifies which is appropriate for the incident class and how to execute it.

Post-incident actions. Documentation requirements, post-mortem expectations, evidence preservation, regulatory notification if applicable. At tier 4 and above, the post-incident phase includes considerations that standard service post-mortems do not — particularly the question of whether the incident represents a tier-violation that requires governance escalation beyond the immediate response.

Ownership at 2am

Returning to the framing question. The runbook is useful only if there is a named human who owns the response. Three structural considerations.

The agent owner is not the agent builder. The team that built the agent typically does not own its production behavior, because production responsibility shifts to operations after deployment. The runbook must name a specific operations owner, not a builder, as the on-call escalation target.

The agent owner has cross-functional dependencies. When an agentic AI incident fires, the operations owner often cannot fix the issue alone — they need engineering support to investigate, legal support to assess regulatory implications, vendor support to address vendor-side issues, and product support to manage customer communication. The runbook must name these supporting roles explicitly, with paging paths for each.

The agent owner needs decision authority. Containment actions — dropping the agent to a lower tier, suspending the deployment, triggering circuit breakers — have business consequences. The operations owner must have either the authority to make these decisions independently or a clear escalation path to a named human who does. A runbook that requires consensus before containment is a runbook that fails the 2am test.

The org-design implications of this — where the operations owner sits, how they are funded, what their relationship is to engineering and to the CAIO function where one exists — are treated in detail in Module 5. The chapter here notes only that the role must be named and resourced, and that organizations operating at tier 4 and above without a named operations owner for each agentic deployment are operating without the response capability the deployment requires.

The post-mortem discipline

A specific aspect of incident response worth treating separately. Post-mortems for agentic AI incidents have requirements that standard service post-mortems do not.

The post-mortem must reconstruct the session-level events in detail. This is not optional. Agentic AI incidents often have non-obvious contributing factors that surface only in the full session reconstruction; a post-mortem that summarizes the session in a paragraph misses what the post-mortem is for. The reconstruction should include the agent's reasoning trace, the retrieved context, the tool invocations, and the downstream effects, in chronological order with annotations on the points where the failure became inevitable.

The post-mortem must distinguish performance failure from tier-violation failure. A wrong output that was within the agent's intended behavior space is a performance issue; a wrong output that the agent should not have been producing at all is a tier issue. The post-mortem must reach a determination on this question, because the remediation differs: performance issues are addressed through model and prompt improvement, tier issues are addressed through architectural change.

The post-mortem must produce specific commitments with named owners and dates. Generic commitments — *we will improve our monitoring* — are not commitments. Specific commitments — *the integrity dashboard will add a new metric for this failure type by this date, owned by this person* — are. The discipline is inherited from blameless post-mortems in SRE practice, applied to the agentic dimension.

The post-mortem at tier 4 and above should be reviewed by the cadence governance from Module 1 — the operational owner for the immediate disposition, the executive level for systemic patterns. Patterns across post-mortems are themselves governance signals; an organization with three post-mortems in a quarter all attributing failures to context-retrieval issues has discovered something the individual post-mortems could not surface.

Monday artifacts

CTO artifact: Incident response architecture decision record

A document that names the runbook owner, the supporting role network, the containment authority chain, and the specific tier-violation classification matrix for the deployment. Reviewed quarterly. Updated whenever the deployment's architecture changes or whenever an incident reveals a gap in the existing runbook.

The required sections. *Operations owner* — named human, with reporting line and authority scope. *Supporting roles* — engineering, legal, compliance, vendor, product, with named contacts for each. *Containment authority* — what decisions the operations owner can make independently and what requires escalation. *Tier-violation classification* — the matrix that distinguishes performance incidents from tier incidents for this deployment specifically. *Runbook location* — where the actual runbook lives, who maintains it, and how it is tested. *Last tested date* — when the runbook was last exercised in a tabletop or live drill, with the next test date scheduled.

The ADR is a one-page document. The runbook itself is longer, lives in operations documentation, and is referenced from the ADR.

Architect artifact: Logging schema and reversibility primitive specification

The implementation-ready specification for the logs and the reversibility primitives. The schemas from Section 1 of this chapter, expanded into actual JSON Schema definitions, with retention policies, access controls, and query patterns specified. The reversibility primitives from Section 2, expanded into the specific retraction protocols, transactional patterns, and compensating-action mappings for the deployment's specific integrations.

The specification should be testable. Each schema field should have a validation rule. Each reversibility primitive should have a test plan. The architect should be able to hand the specification to an engineering team and have the implementation in production within sprints.

A specific note on testing. The reversibility primitives should be exercised periodically in non-production environments, against synthetic incidents that resemble plausible production failures. The testing surfaces gaps that did not appear during initial implementation — primitives that worked against simple test cases but fail against the messier inputs production produces. Quarterly testing is reasonable for tier 4 and 5; monthly for tier 6.

CPO artifact: Incident-readiness portfolio review

Conducted semi-annually. Asks four questions across the deployment portfolio.

What deployments do we operate, at what tiers, and what is each deployment's incident-readiness state? The portfolio inventory plus the readiness assessment.

What gaps exist between the readiness state and the readiness requirement for each deployment's tier? A tier-5 deployment without exception-queue monitoring is under-ready; the gap is a governance liability the CPO owns.

When was each deployment's runbook last tested? Untested runbooks are aspirational rather than operational. A runbook that has not been tested in twelve months should be treated as if it does not exist.

What is the remediation plan for the gaps? Specific commitments, named owners, dates, budget allocation.

The proportionality question from Module 2 applies here. Incident readiness investment should match the risk posture of the deployment. A tier-5 deployment with the incident infrastructure of a tier-3 deployment is a portfolio liability; a tier-3 deployment with tier-5 incident infrastructure is over-invested. The CPO's role is keeping the proportionality calibrated.

Forward references

Module 4 (Procurement and Vendor Evaluation) covers vendor incident-response obligations in detail — what to require contractually about vendor support during agentic AI incidents, including specific SLAs for vendor-side root-cause investigation, evidence cooperation, and remediation. The runbook this chapter specifies typically depends on vendor cooperation that procurement contracts often do not require explicitly.

Module 5 (Organizational Design) covers the operations-owner role in detail, including where it sits in the org chart, how it is funded, and its relationship to engineering, to the CAIO function where one exists, and to the risk function. The chapter takes a position on this that may differ from current standard practice.

Module 6 (Regulatory Posture) covers the regulatory implications of incident handling, particularly the EU AI Act's incident-reporting obligations for high-risk systems (Article 73) and the GDPR's data-breach notification requirements where personal data is involved. The runbook from this chapter must include these notification obligations explicitly; Module 6 specifies what they require.

Module 4: Procurement and Vendor Evaluation

The how. What to ask vendors before signing, what contract language to push for, what verification mechanisms to build, and how all of this differs across the autonomy gradient.

The recommendation, stated first

Most agentic AI procurement contracts in 2026 are written with SaaS-era assumptions that fail when applied to autonomous systems. The default vendor terms over-claim data rights, under-commit to law compliance, and shift accountability to the customer in ways that are inconsistent with the customer's actual regulatory exposure. The 2025 TermScout analysis of AI vendor contracts, as highlighted by Stanford CodeX, found that 92 percent claim data usage rights beyond what is necessary for service delivery — compared to 63 percent in standard SaaS — and only 17 percent commit to compliance with all applicable laws — compared to 36 percent in SaaS. The vendor playbook is more aggressive than what operators tolerate from non-AI software suppliers, and operators who sign agentic AI contracts using their existing SaaS review process are accepting positions worse than their own procurement standards would otherwise permit.

The recommendation is to restructure agentic AI procurement around four specific shifts: from platform access to defined services, from SOC 2 reports to operational audit rights, from uptime SLAs to outcome-based performance warranties, and from one-time onboarding verification to continuous post-deployment monitoring. Each shift has tier-specific provisions; the higher the tier the deployment is intended to occupy, the more aggressive the procurement posture must be.

What the field has already produced

The strongest single contribution to this question is Mayer Brown's February 2026 piece *Contracting for Agentic AI Solutions: Shifting the Model from SaaS to Services*. The piece identifies six specific SaaS contract clauses that fail for agentic AI and proposes business-process-outsourcing-style replacements. The argument is that SaaS contracting was designed for passive tools where the customer is responsible for all consequential decisions; agentic AI is autonomous action where the system is making decisions and taking actions in the world; the contract framework has to follow the system from licensing access to defining services.

This chapter engages with the Mayer Brown framework directly and extends it. The six-clause restructuring — service definition, audit rights, warranties, governance, liability, data ownership — is the right starting point. What it does not yet address is the tier-specific layer: an organization buying a tier-3 deployment requires materially different contract terms from one buying a tier-5 deployment, and the same vendor may be selling both under the same boilerplate. The chapter's contribution is the tier-specific layer above the Mayer Brown framework.

Two other anchors to acknowledge briefly. Morgan Lewis's April 2026 piece confirms the broader market direction: sophisticated buyers are pushing for service-oriented terms, performance-based warranties, governance and audit rights, and outcome-tied liability structures. The General Services Administration's March 2026 proposed clause GSAR 552.239-7001 sets the federal-template benchmark for aggressive procurement language: 72-hour incident reporting, prime contractor liability for downstream commercial AI

Service Providers, government audit and suspension authority, prohibition on training-data use. Private-sector buyers will reference the GSA clause in coming years. Operators looking for the most aggressive defensible procurement posture should treat it as their starting point.

Section 1: Service definition

The SaaS contract defines the service as access to a hosted platform; the customer takes responsibility for what they do with it. This framing collapses for agentic AI because the system is taking actions, not providing access. The contract must define what the agent does, what it is permitted to do, and what it is not.

The Mayer Brown framing names this as *delegation of authority* — the contract specifies what tasks the provider agrees to complete using AI agents, with explicit boundaries on what the agent can and cannot do — combined with *policy guardrails* that specify how the agents can operate, including mandatory escalation triggers for human-in-the-loop approval. This is the right structural move. The chapter extends it with tier-specific guidance.

At tier 1 and 2 (suggestion and draft-and-edit). The service definition is closer to traditional SaaS — the vendor provides access to a copilot capability; the customer is responsible for how their users employ it. The procurement-specific concerns are user training obligations on the vendor side and behavioral monitoring rights on the customer side. The contract should include vendor commitments to training material that explicitly distinguishes tier-1 from tier-2 operation, so that customer users understand which tier they are operating at and the governance posture follows.

At tier 3 (approve-and-send). Service definition expands to include the approval interface specifications. The contract should require the vendor to implement the review interface in ways that support meaningful approval — surfacing the underlying data, the agent's reasoning to the extent exposed, and the alternatives the agent considered. The vendor should commit to logging the approval-event timestamp and the presentation timestamp, because without those the customer cannot demonstrate compliance with meaningful-human-involvement doctrine. Vendors who refuse this commitment are selling a tier-3 product that operates as tier 4 under any inspection.

At tier 4 (supervised execution). Service definition becomes the BPO-style task definition Mayer Brown describes. The contract specifies what categories of action the agent can take, what targets it can act against, what magnitudes are permitted, and what categories require escalation. The contract should include explicit prohibitions — actions the agent cannot take regardless of the agent's reasoning — and explicit thresholds — magnitudes above which actions must escalate. These are the contractual translation of the boundary architecture from Module 1's tier-5 chapter, applied to a tier-4 deployment with looser boundaries.

At tier 5 (bounded autonomy). Service definition includes the boundary architecture as a contractual artifact. The boundary specification is not an internal vendor document; it is contractual deliverable, with versioning, change-control obligations, and customer review rights for material changes. The contract should require the vendor to provide the customer with the current boundary specification, the change history, the approval chain for changes, and notice of proposed changes before implementation.

At tier 6 (self-modifying autonomy). Service definition has to address the policy-versioning question directly. The contract should specify what triggers a policy update, what evaluation precedes deployment of the updated policy, what monitoring follows, and what rollback rights the customer has. Vendors operating at tier 6 without contractual commitments on policy versioning are asking the customer to accept architectural risk the customer cannot assess.

Illustrative service-definition language for tier 4 (jurisdiction-dependent, work with counsel):

> *Provider's service consists of operating an autonomous AI agent ("Agent") on Customer's behalf to perform the following actions ("Permitted Actions"): [enumerate]. The Agent shall not perform any action outside the Permitted Actions, including but not limited to: [enumerate*

> *prohibited actions]. The Agent shall escalate to Customer-designated human reviewers any action where: [enumerate escalation triggers]. Provider shall log every action taken or considered by the Agent, including the action attempted, the determination whether the action was permitted, the reasoning underlying the determination where exposed by the model, and the disposition.*

The specific enumerations are use-case-dependent. The structure is generalizable.

Section 2: Audit rights and verification

The SaaS audit clause is typically limited to providing a SOC 2 report; some agreements give the vendor the right to audit the customer for license overages. This is inadequate for agentic AI because the SOC 2 covers the vendor's security posture and not the agent's behavior, and the audit-overage flow is reversed from where the audit value lies.

The Mayer Brown framing proposes broader transparency rights — technical (access to AI decision logs to answer "why did it do that?") and operational (formal assessment of agent performance against agentic SLAs). The chapter extends with two specific verification mechanisms.

The tier-claim verification mechanism. The contract should give the customer the right to verify that the deployed system operates at the tier the vendor claimed. The verification involves running the diagnostic from Module 1 against the actual production deployment — not the demo, not the sandbox, the production system. The contract should require the vendor to provide the data necessary to compute the tier-integrity metrics from Module 2: the diff log at tier 2, the reviewer-latency log at tier 3, the action-trace and propagation-latency data at tier 4, the boundary-evaluation log at tier 5, the policy-versioning data at tier 6. Vendors who refuse to provide this data are selling a system that cannot be verified against its own marketing claims.

The post-deployment monitoring obligation. EU AI Act Article 72 requires providers of high-risk AI systems to establish post-market monitoring throughout the system's lifetime. The contract should mirror this obligation regardless of the EU AI Act's applicability to the specific deployment, because the monitoring is operationally necessary even where it is not regulatorily required. The contract should specify what the vendor monitors, what data the vendor provides to the customer on what cadence, what triggers vendor-side investigation, and what triggers escalation to the customer.

Audit cadence. Quarterly is reasonable for tier 4 and below; monthly for tier 5; continuous-with-quarterly-deep-review for tier 6. The audit cadence should match the tier-integrity metric cadence from Module 2, because the same data feeds both.

The earn-back trap. A specific procurement gotcha worth naming explicitly. Some vendor SLA structures include "earn-back" provisions that allow the vendor to recover service credits paid for SLA defaults if the vendor performs above the standard for a defined period afterward. This provision undermines the purpose of service credits — the vendor experiences no net cost for SLA failure as long as subsequent performance meets baseline. The contract should exclude earn-back provisions; service credits should be paid against failures and not subject to recovery through subsequent performance.

Illustrative audit-rights language for tier 4 and above:

> *Customer shall have the right, upon reasonable notice, to audit Provider's performance of the Service against the contractual commitments specified herein, including the right to access (i) Agent decision logs sufficient to reconstruct the reasoning underlying any specific Agent action, (ii) tier-integrity metrics as defined in [reference], computed against actual production data over the audit period, and (iii) Provider's post-deployment monitoring reports as required under Section [post-deployment monitoring]. Provider shall cooperate with audits conducted by Customer or by Customer's third-party representative, and shall not condition such cooperation on Provider's prior approval of the auditor.*

Section 3: Performance warranties and SLAs

The traditional SLA framework — uptime guarantees, latency thresholds, response-time commitments — was designed for systems whose value is availability. Agentic AI systems are valuable for outcomes, not availability. An agent that is available 99.9 percent of the time but produces wrong outputs is operationally worse than an agent that is available less often but produces correct ones.

The shift from technical-metric SLAs to outcome-based SLAs is the second-largest restructuring in agentic AI procurement after the service-definition shift. The chapter recommends operators push for outcome-based performance warranties wherever the outcomes can be defined cleanly enough to measure.

At tier 1 and 2. Outcome-based warranties are difficult because the human is in the production loop and outcome attribution is ambiguous. The relevant warranties are usually about platform behavior (the suggestion surface is available, the model produces output within latency targets) and about model quality (accuracy on representative tasks above defined thresholds). The CLEAR and CLASSic frameworks from Module 2 provide useful warranty dimensions.

At tier 3. The relevant outcomes are typically the same as tier 1 and 2 plus the approval workflow performance (approvals process within acceptable latency, the interface surfaces required information). The vendor-side commitment is to support the customer's reviewer-effectiveness measurement, not to guarantee the reviewer's effectiveness, which is the customer's responsibility.

At tier 4. Outcome-based warranties become primary. The agent is acting in the world; what matters is whether its actions produce the outcomes the customer is paying for. For an AISDR deployment, the outcome is meeting-booked, qualified-lead-generated, or revenue-attributed. For a customer service agent, the outcome is case-resolved-without-human-handoff. Sierra AI's pure outcome-based pricing model — payment only for successfully resolved cases — is the existence proof that this structure works at scale. Operators should push for vendor SLAs tied to these outcomes, with service credits when outcomes fall below thresholds.

At tier 5. Outcome-based warranties continue to apply, with the addition of boundary-effectiveness warranties. The vendor should warrant that the boundary architecture catches the cases it was designed to catch, with measurable boundary-effectiveness metrics and remedies when the boundaries fail to engage as intended.

At tier 6. Performance warranties have to address policy-stability dimensions. The vendor should warrant that policy updates preserve specified behavioral characteristics, with measurable drift thresholds and remedies when drift exceeds warranted bounds. This is the hardest warranty to draft because the field has not yet standardized what acceptable drift looks like; the chapter's recommendation is that operators not deploy at tier 6 without this warranty in place, even if it requires negotiating bespoke language for each deployment.

The "what counts as an outcome" definition. The most common procurement failure in outcome-based contracting is ambiguous outcome definition. A vendor and customer who both signed a contract guaranteeing "qualified leads" can disagree productively for years about what counts as qualified. The contract should define outcomes operationally — what data confirms the outcome, who measures it, what dispute-resolution applies when measurements disagree. This is hard to draft and expensive to negotiate; it is also the work that makes outcome-based contracts enforceable rather than aspirational.

Section 4: Pricing models

The 2026 pricing landscape is unsettled. Salesforce runs three pricing models for Agentforce simultaneously — per-conversation, Flex Credits per-action, per-user — plus the Agentic Enterprise License Agreement (AELA) for full-scale deployment. Microsoft's M365 E7 (announced March 2026) bundles agentic capability into a $99/user/month tier with embedded Copilot Credit metering for agent activity. Sierra AI runs pure outcome-based pricing and reached $150 million ARR by early 2026. Zendesk shifted to outcome-based at $1.50 per automated resolution. The market has not converged.

The chapter recommends a tier-specific pricing posture rather than a single model.

Per-seat pricing — pay per user with access to the agent capability. Appropriate for tier 1 and tier 2 deployments where the agent augments human work and the human is the natural unit of the spend. Inappropriate for tier 4 and above where the agent acts independently of any specific user; per-seat pricing in autonomous deployments either over-charges (paying per seat for users whose work is mostly automated) or under-charges (paying per seat for unbounded autonomous activity).

Consumption-based pricing — pay per action, per token, per conversation. The model most enterprise software vendors experimented with in 2024 and 2025; the model CFOs systematically pushed back against because of unpredictability. Appropriate for early experimentation phases where the operator wants to constrain spend by constraining usage. Inappropriate as the long-term model for production deployments because budget unpredictability suppresses adoption.

Agentic Enterprise License Agreement (AELA) — flat fee for unlimited agentic activity over a multi-year term. Appropriate for mature deployments at tier 4 and above where the operator is committed to the platform and wants budget predictability. Inappropriate as a starting position because of the lock-in implications; Constellation Research's prediction that AELA renewals will convert to defined-quantity contracts at end of term — compounded by projected 6-15 percent annual increases above inflation (Gartner) — is a real procurement risk that should be priced into the original AELA decision. The Adecco-Salesforce AELA through 2027 is the live case study.

Outcome-based pricing — pay only for successful outcomes (resolved cases, booked meetings, completed transactions). Appropriate for narrowly-scoped agents where the outcome can be defined cleanly and measured unambiguously. Sierra's resolution-based pricing for customer service is the cleanest example. Inappropriate for broader agent deployments where outcomes are ambiguous or depend on factors outside the vendor's control.

Hybrid models — combinations of the above. Salesforce's three-models-running-simultaneously is the most visible example; many enterprise procurement teams are negotiating bespoke combinations. The chapter's position: hybrid models are appropriate when no single model fits the deployment cleanly, but the contract must be explicit about which pricing applies to which activity and what triggers transition between pricing models. Vague hybrid contracts are renewal-time disaster.

My recommendation by tier. Tier 1 and 2: per-seat, with vendor commitments to seat-cost stability across renewals. Tier 3: per-seat, possibly with consumption overlay for high-volume approval activity. Tier 4: outcome-based where outcomes are clean; AELA where the operator is committed to platform. Tier 5: AELA, with explicit attention to renewal-time conversion mechanics. Tier 6: bespoke; the deployments are rare enough that pricing is negotiated case by case.

The "unlimited" trap. AELA marketing typically uses "unlimited" framing. The reality, per Constellation Research and multiple other sources, is that "unlimited" is the headline number; data fees, connection fees, integration fees, and specific feature surcharges sit underneath. The contract should require the vendor to disclose all auxiliary fees that apply during the AELA term and commit to caps on auxiliary fee increases at renewal. Unlimited that is unlimited at the action level but uncapped at the data-fee level is not unlimited; it is a metered system with the meter hidden under the headline price.

Section 5: Post-deployment monitoring obligations

The pre-deployment evaluation that procurement contracts have historically focused on is necessary but insufficient. The 37-percent lab-to-production gap from AWS research, the NIST CAISI March 2026 report on monitoring challenges, and the EU AI Act Article 72 post-market monitoring requirement all point in the same direction: the contract must require continuous post-deployment verification, not one-time onboarding verification.

The contract should specify what the vendor monitors, what data the vendor provides to the customer on what cadence, what triggers vendor-side investigation, and what triggers customer escalation rights.

Vendor monitoring obligations at minimum should include: tier-integrity metrics for the deployment's claimed tier (per Module 2), performance metrics against the warranted SLAs (per Section 3 of this chapter), policy-version evaluation results at tier 6 (against drift thresholds), and incident occurrence with defined severity classification.

Reporting cadence should match the cadence from Module 2: weekly for behavioral metrics, monthly for trend reports, quarterly for performance reviews. The contract should specify report content, format, and delivery channel.

Customer access to underlying data is the load-bearing provision. Vendor-summary reports are useful but insufficient — the customer must have access to the underlying logs to support independent verification. The contract should grant the customer (and the customer's representatives) access to decision logs, policy-version data, action traces, and other underlying data on reasonable request. Vendors who limit this access to summary reports are limiting the customer's ability to demonstrate compliance with the customer's own regulatory obligations.

Material change notification — the vendor must notify the customer of material changes to the deployed system within a defined window. "Material" should be defined: model version changes, prompt template updates, tool integration changes, boundary specification changes at tier 5, policy updates at tier 6. The notification window should be 30 days or less for non-urgent changes; immediate for changes that affect security or compliance posture.

Section 6: Vendor-side incident response

Module 3 specified the operator-side runbooks for agentic AI incidents. The procurement contract is where the vendor-side obligations during those incidents become enforceable.

Vendor incident response commitments at tier 4 and above should include: specific SLAs for vendor-side investigation initiation (typically within hours of customer escalation), evidence cooperation (vendor commits to preserve and provide all logs, traces, and system state relevant to incident reconstruction), root-cause analysis delivery within a defined window, and remediation commitments including specific timelines for fixing identified issues.

The 72-hour standard. The GSA proposed clause's 72-hour incident reporting requirement is becoming a benchmark. Operators should consider negotiating equivalent vendor commitments — the vendor reports incidents detected on the vendor side to the customer within 72 hours, regardless of whether the customer has independently noticed them. This commitment is unfamiliar to vendors but is consistent with where regulatory requirements are heading.

Cooperation with customer post-mortems. The post-mortem discipline from Module 3 requires session-level reconstruction. That reconstruction is impossible without vendor cooperation if the incident touched vendor-side components. The contract should require vendor participation in customer-led post-mortems, including provision of vendor-side logs and access to vendor engineers for the reconstruction.

Liability allocation for incidents. The most contested provision in agentic AI contracts. The vendor's standard position is liability cap at fees paid; the operator's position should be liability proportionate to the harm the agent caused, given the vendor's commitments about what the agent would and would not do. The chapter's position: liability cap at fees paid is acceptable for tier 1 and 2; not acceptable for tier 4 and above. At higher tiers, the vendor is operating an autonomous system on the customer's behalf, and liability should reflect the consequences of system failure, not just the cost of the service.

Section 7: Data rights, IP, and lock-in

The TermScout finding — 92 percent of AI contracts claim data usage rights beyond what is necessary for service delivery — names the dominant procurement failure. Operators who sign vendor-default data clauses are giving the vendor rights to use customer data for training, for competitive intelligence, and for purposes the customer never authorized.

Data rights to push back on:

The contract should restrict vendor data use to what is necessary for service delivery. The vendor should not be permitted to use customer data for training models, for benchmarking other customers, for competitive intelligence, or for any other purpose without explicit per-use authorization. The vendor's standard contract often includes a broad license to "improve our services" that effectively permits training; this should be removed.

The contract should specify data retention limits. The vendor should retain customer data only for the duration necessary to provide the service, plus a defined post-termination window for transition. Indefinite retention should not be acceptable.

The contract should include data portability obligations. On termination, the vendor should provide customer data in a defined format within a defined window, with no charge for the export. Data lock-in through proprietary formats or excessive export charges is increasingly a procurement-rejection condition.

IP rights for outputs. The contract should specify ownership of agent outputs. The customer should typically own the outputs; the vendor should warrant that the outputs do not infringe third-party rights, with indemnification for infringement claims. Some vendors (Google, Microsoft, IBM) provide IP indemnification on AI outputs as standard; many do not. Operators should not accept agentic AI contracts without output-IP warranties at tier 2 and above.

The lock-in question. Kai Waehner's 2026 enterprise agentic AI landscape mapped vendors on two dimensions: trust and lock-in. The lock-in dimension is structurally important and procurement-addressable. Lock-in compounds at every layer of the stack — model, orchestration, integration, data, workflow. The contract should require the vendor to support customer flexibility on the dimensions that matter most to the operator's deployment.

Specific lock-in-mitigation provisions: model portability (customer can substitute alternative model providers without losing functionality); orchestration neutrality (the agent can be reconfigured to use alternative tool integrations without vendor approval); data portability (covered above); and abstraction commitments (the vendor maintains stable interfaces that survive vendor-side architectural changes).

The geopolitical-data-jurisdiction layer. A specific concern that has become procurement-relevant in 2026 and will remain so. Where is the customer's data processed? Under which jurisdiction does the vendor operate? What happens to the data if that jurisdiction changes? European enterprises are particularly attentive to these questions; US enterprises increasingly so as data sovereignty regulations evolve. The contract should specify processing locations, jurisdictional commitments, and notice obligations for material jurisdictional changes. This is fast-moving regulatory territory; operators should consult specialist counsel for high-stakes deployments rather than relying on standard contract language.

Section 8: Termination and exit

The contract that does not explain how the relationship ends has not been adequately negotiated. Agentic AI termination provisions raise concerns that standard SaaS termination does not.

Termination triggers. Beyond standard breach-and-notice, the contract should include termination rights for: SLA failures above defined thresholds, material adverse changes in the vendor's regulatory posture, material security incidents on the vendor side, and material changes to the underlying model that affect performance against warranted SLAs. The customer should also have for-convenience termination rights with reasonable notice; agentic AI relationships are not relationships customers should be locked into indefinitely.

Exit support obligations. On termination, the vendor must provide: data export in defined formats within a defined window, configuration export sufficient to support migration to alternative platforms, and transition support for a defined period. Vendors increasingly include transition-services obligations as standard; operators should require them.

Post-termination data destruction. The vendor must destroy customer data after the transition window, with certification of destruction. This should include data the vendor used for derived models or benchmarks; the destruction obligation is broader than the source data alone.

The 11x lesson. Customers reported in TechCrunch's March 2025 reporting on 11x that they were unable to exercise contractual break clauses despite successfully negotiating them. The procurement-side lesson: break clauses are only as enforceable as their operational specifics. The contract should specify the procedure for invoking the break clause, the timeline for vendor response, and the remedies if the vendor fails to honor the clause. Vague break clauses without operational specificity are aspirational, not enforceable.

Monday artifacts

CTO artifact: Procurement decision record

A document, completed before contract signature, that captures the architectural choices the contract reflects. The required sections.

The deployment. What is being procured, at what tier, for what business purpose. Cross-references the architecture decision record from Module 1.

The pricing model and its rationale. Why this pricing model rather than alternatives, what risks it shifts to whom, what protective clauses address those risks.

The audit and verification posture. What the contract gives the customer to verify that the deployed system operates at its claimed tier; what data the vendor provides on what cadence; what triggers customer escalation.

The exit posture. What termination rights the contract provides; what exit support obligations apply; what data destruction obligations apply post-termination.

The unverified concessions. What positions the customer accepted that the customer would not have accepted in standard SaaS, with the rationale for accepting them. This is the integrity-of-the-record provision; if the customer accepted broad data rights or capped liability, the document names it explicitly so the choice is examined rather than buried.

Owner. One human, the CTO or designee, who carries the procurement decision.

Review date. Annually, or at any material change in the deployment's tier or use case.

Architect artifact: Vendor evaluation question set

A structured set of questions to ask vendors during the evaluation phase, organized by the four shifts: service definition, audit rights, performance warranties, post-deployment monitoring. The question set is tier-specific; questions for a tier-3 deployment differ from questions for a tier-5 deployment.

For each question, the architect captures the vendor's answer and an assessment: whether the answer is acceptable, whether it requires negotiation, whether it disqualifies the vendor.

Disqualifying answers exist at every tier and operators should be willing to act on them. A vendor who refuses audit rights for tier-integrity verification is selling a system that cannot be verified against marketing claims; that vendor should be eliminated from consideration regardless of price competitiveness. A vendor who refuses to commit to data restriction (no use of customer data for training) is offering a contract significantly worse than standard SaaS; that vendor's pricing must compensate for the data exposure or the vendor should be rejected.

The question set is the architect's instrument for surfacing these issues during evaluation rather than after signature. The questions should be sent to vendors as part of the formal RFP/RFI process, with answers required in writing, so that the answers become part of the evaluation record and ultimately the contractual representations.

CPO artifact: Procurement portfolio governance review

Conducted semi-annually. Asks four questions across the procurement portfolio.

What agentic AI vendor relationships do we have, at what tiers, on what pricing models? The portfolio inventory.

What contractual posture does each relationship reflect? The audit summary across the portfolio — which contracts have full audit rights, which have outcome-based SLAs, which have post-deployment monitoring obligations, which do not.

Where are the gaps between contractual posture and operational risk? A tier-5 deployment with tier-3 contractual posture is under-protected; the gap is the procurement liability the CPO owns.

What is the remediation plan? Renegotiation at next renewal, supplemental agreements, vendor replacement — specific commitments with named owners and dates.

The proportionality theme from Modules 5 and 6 continues here. Procurement investment should match the risk posture of the deployment. A tier-5 deployment with the contractual protections of a tier-3 deployment is a portfolio liability that the CPO must address; a tier-3 deployment with the procurement overhead of a tier-5 deployment is over-invested and should be simplified.

Forward references

Module 5 (Organizational Design) covers procurement function structure — where AI procurement sits in the organization, who owns the vendor evaluation, how the work coordinates with legal, risk, and engineering. The procurement function is often orphaned in organizations where AI cuts across traditional procurement boundaries; Module 5 addresses this directly.

Module 6 (Regulatory Posture) covers the regulatory frameworks that procurement contracts must enable compliance with. EU AI Act Article 72 on post-market monitoring, Article 73 on serious incident reporting, GDPR Article 22 on meaningful human involvement, sector-specific rules — the contract must provide the customer with the rights and data necessary to meet the customer's own regulatory obligations. Module 6 specifies what the regulations require; this chapter has specified what the contract must provide to enable compliance.

Module 5: Organizational Design for Cadence Governance

The how. Who specifically owns what, where they sit, how they coordinate across functions, and what scale-appropriate governance looks like at enterprise, mid-sized, and startup organizations.

The recommendation, stated first

Most enterprise agentic AI deployments fail at the organizational layer rather than the technical layer. The architecture is sound; the governance documents exist; the eval infrastructure has been built; the procurement contract was negotiated competently; the regulatory posture was defensible at deployment. And yet the deployment drifts because no specific human owns the cadence, no escalation path is operational, and the cross-functional coordination that the deployment requires is everyone's responsibility and therefore no one's.

The recommendation is to make ownership explicit at every level of the cadence governance from Module 1, with named humans, defined authority, and resourced capacity. Each tier-integrity metric from Module 2, each runbook from Module 3, each procurement decision from Module 4, and each regulatory obligation from Module 6 needs an identified human owner who can be paged at 2am and an identified escalation path that produces decisions rather than committees. The CAIO question that dominates current AI governance discourse is downstream of this — whether or not your organization needs a

Chief AI Officer depends on what ownership structure you can sustain without one, and the appointment is not a substitute for the underlying ownership architecture.

This chapter is the synthesis chapter for Part III. The previous four chapters named specific roles that need to exist; this chapter resolves them into a coherent organizational map and engages critically with the CAIO discourse rather than uncritically endorsing it.

What the field has already produced

The dominant framework in current discourse is the Chief AI Officer role. *As of April 2026*, IBM's 2026 study reports 76% of organizations have a CAIO in 2026, up from 26% in 2025; roughly 40% of Fortune 500 companies are expected to have appointed one by year-end; recruitment for the role tripled over the past five years. CAIOs report to CEO (43%), CTO/CIO (35%), or COO (12%). More than half report directly to CEO or board. The role is real, it is being adopted at scale, and the chapter must engage with it.

The CIO industry publication's February 2026 critique provides the most thoughtful counter-position. The argument: "the long-term success of this role may reduce the need for its existence... when AI becomes embedded and routine, the title may matter less than the capabilities and governance structures it helped establish." The piece names a specific pattern — organizations absorbing CAIO responsibilities into existing leadership as AI normalizes. This is not a prediction; it is a documented organizational pattern.

The dominant practitioner consensus in AI governance writing — across vendor publications, consultancy frameworks, and industry analyst reports — converges on the same structural answer: AI governance fails when left to a single function and succeeds when ownership is distributed across product, engineering, legal, risk, and operations with explicit accountability allocation, typically through a RACI structure or equivalent.

The minimum viable governance (MVG) framework, articulated by Ajay Pundhir at AI Exponent and adopted across the startup-advisory community, provides a 90-day on-ramp adapted from NIST AI RMF for organizations that cannot sustain enterprise-scale governance infrastructure. Four pillars: Govern, Map, Measure, Manage. Anchored on a 1-Page AI Ethics Charter with explicit "red lines" and core commitments. The argument: comprehensive governance frameworks fail at startup scale because they "demand maturity you don't have yet," and the alternative is not no governance but structurally appropriate governance.

This chapter engages with all three streams — the CAIO discourse, the cross-functional RACI patterns, and the MVG framework — and synthesizes them into a tier-aware, scale-aware organizational design.

Section 1: The cadence-to-role mapping

Module 1 named four cadence levels: weekly, monthly, quarterly, semi-annual. Each level has a different temporal horizon, a different scope of decision, and a different appropriate owner. The chapter's first task is to map these levels to specific organizational roles.

Weekly: behavioral metrics review at the operational team level. The team that operates the deployment reviews the tier-integrity metrics from Module 2 each week. Edit ratios at tier 2, reviewer latencies at tier 3, propagation latencies at tier 4, exception-handling times at tier 5, policy-drift signals at tier 6. Fifteen minutes; one question — *is the trajectory consistent with the tier we committed to?* — with escalation when the answer is no.

The named owner for the weekly review is the *deployment operations lead*. This is typically a senior engineer or technical product manager with day-to-day operational responsibility for the deployment. The role is not "AI governance" in the abstract; it is "this specific deployment's operational owner." The deployment operations lead is the person who would be paged if the deployment had a production incident at 2am, and the person who can authorize containment actions including dropping the deployment to a lower tier when needed.

For organizations operating multiple deployments, each deployment has its own operations lead. The leads coordinate through a shared forum at the monthly cadence rather than at the weekly level. Trying to centralize weekly reviews across multiple deployments fails because the reviews are too granular for cross-deployment attention.

Monthly: trend and exception review at the architecture decision record owner level. The named owner of the deployment's ADR — typically the architect who designed the deployment or the senior engineer who currently maintains the architecture — examines the four-week trend on integrity metrics, the exception patterns from operational logs, and any incidents from the period. Thirty minutes; produces specific actions.

The ADR owner role is distinct from the operations lead role. The operations lead is responsible for the deployment running today; the ADR owner is responsible for the deployment's architectural commitments staying coherent over time. In smaller organizations these may be the same person; in larger organizations they should be different people, because the temporal horizons of weekly operational concerns and monthly architectural concerns require different attention patterns.

Quarterly: governance review at the executive level. The CTO, the CPO (or equivalent), the senior risk function, and the General Counsel meet to look at the integrity dashboards across the deployment portfolio plus the red-team probe results, the sampled output reviews, and any incidents. Sixty minutes; produces decisions about deployment continuation, tier-claim adjustment, vendor performance, and regulatory posture.

The named owner for the quarterly review is whoever convenes it. In organizations with a CAIO, the CAIO convenes; in organizations without, the role typically falls to the CTO with explicit cross-functional invitation. The convening function matters — without a named convener, quarterly reviews become events that get scheduled in principle and skipped in practice.

Semi-annual: strategic review at the board or audit committee level. The senior governance forum examines the integrity dashboards across the deployment portfolio, the cumulative regulatory exposure, and the proportionality of governance investment to deployment scale. Ninety minutes, twice per year. Addresses portfolio-level questions about whether the organization is operating at the right tiers.

The named owner for the semi-annual review is the CEO or CEO-designee, with the board or audit committee as the audience. In organizations with a CAIO, the CAIO presents; in organizations without, the CTO presents with explicit support from the General Counsel and CRO. The semi-annual review is the level at which strategic AI agenda decisions get made, and it requires the seniority to support those decisions.

A summary mapping for the typical enterprise deployment:

- Weekly: deployment operations lead (per deployment)
- Monthly: ADR owner (per deployment)
- Quarterly: CTO + CAIO (where one exists) + CPO + CRO + GC
- Semi-annual: CEO/CAIO presenting to board

This mapping assumes a deployment is significant enough to warrant the full cadence. For lower-tier deployments (tier 1, tier 2), the cadence can compress — monthly reviews instead of weekly, quarterly instead of monthly, with semi-annual continuing as the senior forum. The proportionality theme from previous chapters continues: cadence investment should match tier risk.

Section 2: The integration question — CAIO and alternatives

The CAIO discourse is partly responding to a real organizational need and partly responding to board-level signaling pressure. The chapter takes a position on which is which.

The real organizational need is integration. Agentic AI cuts across CTO (architecture, infrastructure), CDO (data quality, lineage), CISO (model security, threat modeling), CRO (risk frameworks, regulatory exposure), General Counsel (legal interpretation, regulator engagement), and operating functions (specific deployment ownership). No single existing role naturally covers the full surface. In organizations large enough that these are distinct people with distinct reporting lines, the integration function is genuinely orphaned, and a CAIO who reports to the CEO or board can fill the gap.

The board-level signaling pressure is what produces the CAIO discourse's tendency toward uncritical endorsement. Boards want reassurance that leadership "understands what is coming"; investors want signals that the company will not be left behind. Appointing a CAIO is a visible response to both pressures even when the appointment does not change the underlying organizational structure. *As of April 2026*, the appointment-as-signaling pattern is widespread enough that operators should distinguish it from appointment-as-genuine-integration when deciding whether to make the appointment.

The chapter's position on whether the CAIO role is right for a specific organization:

Yes for large enterprises with multiple business units, multi-product portfolios, complex regulatory exposure (multiple jurisdictions, multiple sectors), and existing senior functions (CTO, CDO, CISO, CRO, GC) that are otherwise integrated only at the CEO level. The CAIO provides the coordination layer below the CEO that prevents AI governance from being pulled apart by competing functional priorities. The role pays for itself through faster decision-making and reduced cross-functional friction. IBM's 2026 finding that companies with a CAIO achieved higher returns on AI investments and scaled more AI initiatives than those without is consistent with this view.

Probably not for mid-sized organizations (under $1B revenue, under 5,000 employees, single-product focus). At this scale, the integration function can be discharged by the CTO with explicit cross-functional governance committee support. Appointing a CAIO at this scale tends to isolate AI governance from the operational functions where it should sit, which is the failure mode the CIO

February 2026 critique correctly identifies. The fractional CAIO pattern (treated below) is often the right answer for organizations in this range that need external accountability without full-time appointment.

Definitely not for startups. Startup founders are themselves the integration function. Appointing a CAIO before the organization has the differentiated functions that need integration is making a structural mistake — the role will not have the coordinating relationships to do its job. Startups should focus on minimum viable governance (treated in Section 5) rather than role appointment.

Possibly transitional even at large enterprises. The CIO February 2026 critique's observation that absorbing CAIO responsibilities into existing leadership as AI normalized may be the correct long-term equilibrium for many organizations. The CAIO role solves the integration problem during the period when AI is novel and uncoordinated; once governance maturity is established, the function can be distributed back into existing roles. Operators appointing CAIOs today should plan for the role's possible evolution rather than treating the appointment as permanent.

The CDAO question. Some organizations combine the Chief Data Officer and Chief AI Officer roles into a single Chief Data and AI Officer (CDAO). The combination is sensible where data and AI are strategically coupled and the same governance posture applies to both. It is problematic where data governance and AI governance have different risk profiles requiring different specialized attention — particularly in regulated industries where data privacy compliance and AI compliance follow different doctrinal paths. The chapter's recommendation: combine when the underlying functions can be effectively governed by a single person with both backgrounds, separate when the combination produces a role that cannot specialize in either domain adequately.

The General Counsel relationship. The most-likely-to-conflict relationship in the AI governance organization. The GC owns regulatory interpretation and regulator engagement; the CAIO owns AI strategy and operational governance. The conflict point: when regulatory interpretation says one thing and operational pressure says another, who decides. The chapter's recommendation: explicit charter language that the GC's regulatory interpretation governs absent specific exception, with exception decisions made at the CEO level rather than between CAIO and GC. This is unambiguous and avoids the failure mode where AI deployment continues despite legal concerns because the conflict was never escalated.

Section 3: The behavioral-monitoring orphan

A specific function that does not naturally belong to any existing role: behavioral monitoring of agentic AI deployments. The tier-integrity metrics from Module 2, the drift detection from the same chapter, the patterns surfacing in incident post-mortems from Module 3 — these all require ongoing attention from someone whose job is *watching how the system behaves over time and detecting drift before it becomes incident.*

The function does not naturally belong to engineering. Engineering owns the system, not the user behavior, and engineers' attention naturally concentrates on system performance rather than tier-claim integrity. The integrity dashboard sits in the engineering org because that is where the data is, but engineering ownership of the function tends to produce dashboards that exist without being watched.

The function does not naturally belong to risk. Risk frameworks think in terms of identified risks rather than tier-claim verification; the risk function tends to ask "what could go wrong" rather than "is this still operating at the tier we claimed." When risk owns behavioral monitoring, the behavioral metrics get translated into risk language and lose the operational specificity that makes them useful as drift-detection instruments.

The function does not naturally belong to product. Product optimizes for adoption and user satisfaction; behavioral metrics are more likely to surface adoption-friction issues from the product perspective than tier-integrity issues from the governance perspective. Product ownership of behavioral monitoring tends to weight the metrics toward usability rather than governance.

The function needs a new operational discipline. The chapter's recommendation: a dedicated AI Operations function (sometimes called AIOps, though the term is overloaded) responsible for the behavioral monitoring of all agentic AI deployments. The function reports to the CAIO where one exists, to the CTO where one does not, and the function's specific deliverables are the integrity dashboards, the drift-detection probes, the sampled output reviews, and the cross-deployment pattern analysis.

Specific staffing implications. AI Operations is staffed differently from traditional MLOps or DevOps. The skill profile is closer to an SRE crossed with a data analyst — operational discipline plus the analytical capability to interpret behavioral metrics in context. *As of April 2026*, the role does not have a standardized title; operators are inventing it. The chapter's recommendation: establish the function explicitly with a named lead, regardless of what title the organization eventually uses. The function's existence matters more than its name.

Capacity sizing. For organizations operating fewer than ten agentic AI deployments at tier 4 or above, AI Operations can be a fractional role within engineering or governance. For organizations operating ten or more, it warrants a dedicated function with at least one full-time owner. For organizations operating twenty or more, it warrants a small team with rotation responsibility across deployments.

Section 4: Cross-functional coordination — RACI patterns at each tier

The cadence-to-role mapping in Section 1 establishes who owns what at each cadence level. The tier-specific RACI patterns establish who is consulted, who approves, and who is informed for specific decisions within each cadence.

The patterns differ across tiers because the decisions differ. Specific examples drawn from current operator practice:

At tier 3 (approve-and-send). The dominant cross-functional decision is reviewer-effectiveness governance. Who decides when reviewer latencies have compressed below the meaningful-review threshold and the deployment must be modified? The operational owner is responsible (R) for surfacing the issue; the CAIO or CTO is accountable (A) for the decision; the General Counsel is consulted (C) on regulatory implications under GDPR Article 22; the operating function whose users are doing the review is consulted (C) on operational impact; the CRO is informed (I). Decision authority rests with the CAIO/CTO based on legal input.

At tier 4 (supervised execution). The dominant decision is the dialer-check determination — when does propagation latency on authoritative records cross the threshold that triggers boundary engagement or temporary suspension? Operational owner R, CAIO/CTO A, vendor management consulted (C) where a vendor is involved, business unit affected consulted (C), CRO and GC informed (I).

At tier 5 (bounded autonomy). The dominant decisions are boundary changes — adding a new boundary, relaxing an existing boundary, expanding the agent's permitted actions. The boundary architect is R; the CAIO or CTO is A; the General Counsel is consulted (C) on legal and regulatory implications; the CRO is consulted (C) on risk implications; the operational owner is consulted (C) on operational implications; the affected business units are informed (I). Boundary changes are not engineering decisions; they are governance decisions with engineering implementation.

At tier 6 (self-modifying). The dominant decisions are policy-update approvals — when can the system update its policy based on production feedback. The AI Operations lead is R for surfacing the policy-update events; the CAIO is A for approval; the General Counsel is consulted (C) given the substantial regulatory exposure; the CRO is consulted (C); the operational owner is consulted (C); the board is informed (I) as part of semi-annual review. The chapter's earlier observation stands: most operators should not be at tier 6, and the RACI complexity here is part of the reason.

A specific cross-tier observation. The General Counsel is consulted on most cross-functional AI governance decisions at tier 4 and above. This is not a peripheral consultation; it is a core dependency. Organizations whose GC is not staffed to participate in AI governance at this cadence — who treat AI as an engineering issue with occasional legal review — will discover the GC dependency the hard way during the first significant incident or regulatory inquiry. The chapter's recommendation: invest in GC capacity for AI governance proactively, including potentially a dedicated AI counsel role at large enterprises.

Section 5: Scale-appropriate governance

The organizational design recommendations above assume an enterprise of meaningful scale. Most agentic AI deployments are happening at organizations of varying scale, and the governance design should fit the scale. Three regimes worth treating distinctly.

Enterprise scale (>$1B revenue, >5,000 employees). The full cadence governance from Section 1 applies. The CAIO appointment question (Section 2) is genuinely active. The behavioral-monitoring orphan (Section 3) is a real organizational gap that AI Operations should fill. Cross-functional RACI patterns (Section 4) need explicit documentation. The governance overhead is real but proportionate to deployment scale.

Mid-sized scale ($100M-$1B revenue, 500-5,000 employees). The cadence governance applies in compressed form. Weekly and monthly reviews continue as described; quarterly reviews include CTO, CPO, CRO, GC but typically without dedicated CAIO. The integration function falls to the CTO with explicit governance committee support. Behavioral monitoring is a fractional role within engineering or governance rather than a dedicated function.

Cross-functional RACI exists but is documented at the deployment level rather than as enterprise-wide policy.

The fractional CAIO pattern is often appropriate at this scale. *As of April 2026*, multiple commercial firms (chiefaiofficer.com, AI Smart Ventures, others) offer 90-day fractional CAIO engagements. The pattern provides external accountability and structural discipline without the headcount commitment of full-time appointment. The chapter's view: fractional CAIO is a legitimate steady-state arrangement for many mid-sized organizations, not just a placeholder for eventual full-time appointment.

Startup scale (<$100M revenue, <500 employees). The full cadence governance does not apply. The minimum viable governance (MVG) framework adapted from NIST AI RMF — Govern, Map, Measure, Manage — is the appropriate substitute. Specific anchors:

A 1-Page AI Ethics Charter with named "red lines" (things the company will not do regardless of capability) and core commitments. The charter is operational, not philosophical — it serves as the North Star for product decisions when trade-offs arise.

A simple AI inventory: list of AI systems in use, who owns each, what tier they operate at, what regulatory exposure they have. Reviewed quarterly. The inventory is the foundation; everything else assumes it exists.

A designated owner for AI governance — typically the CTO or VP Engineering at a startup, sometimes the founder if technical. The role is fractional; the responsibility is real. The owner is the named human who would be called if the company received a regulatory inquiry about its AI systems.

Tier discipline in deployment decisions. Startups can and do deploy at higher tiers than their governance maturity supports because the deployment friction is lower. The MVG framework's specific recommendation: pick deployment tiers deliberately, with the constraint that the tier should not exceed what the governance structure can sustain. A startup that cannot run weekly behavioral metric reviews should not deploy at tier 5.

The startup-scale governance is not lower quality; it is structurally different. The MVG framework is genuinely effective at startup scale and genuinely fails at enterprise scale because the integration challenges differ. Operators should match the framework to the scale rather than scaling down enterprise frameworks or scaling up startup frameworks.

Section 6: The transition question

Most readers will not be designing organizational structure from scratch. They will be working with an existing structure that may not match the recommendations in this chapter, and the question is what to change.

The chapter's recommendation: do not attempt comprehensive reorganization. The cost of reorganization is high, the disruption is significant, and the path to a better state typically involves several incremental changes rather than a single restructuring. Specific incremental moves worth prioritizing:

Name the deployment operations lead for every existing tier-3+ deployment. This is the single most effective intervention because most organizations have not done it. The cost is essentially zero (no headcount addition, just role designation); the value is that someone is now accountable for the deployment's behavior at the weekly cadence. If you do nothing else from this chapter, do this.

Establish the quarterly review forum. Sixty minutes, four times per year, with the CTO/CPO/CRO/GC and CAIO where one exists. The forum's primary purpose is to ensure the cross-functional dependencies on AI governance are operational. Forums that did not exist before tend to surface issues that were suppressed by the absence of structure.

Address the behavioral-monitoring orphan. Either name an AI Operations function or assign behavioral monitoring explicitly to engineering or governance. The function exists in your organization whether or not you name it; naming it produces accountability and prevents the metrics from being treated as data nobody owns.

Decide on the CAIO question deliberately. If your organization is at the scale where the CAIO role might be appropriate, make the decision based on the integration analysis from Section 2 rather than based on board pressure or peer-organization signaling. If the decision is yes, follow with a clear charter and reporting structure; if no, follow with explicit allocation of the integration function to the CTO with cross-functional support.

Invest in General Counsel capacity for AI governance. This investment surfaces last in most organizations and should surface earlier. The GC's involvement in tier-4-and-above governance is not optional; the question is whether it is staffed proactively or reactively after the first significant incident.

The transition typically takes 12-18 months at enterprise scale. The chapter's recommendation: sequence the changes so each delivers value independently, rather than waiting for comprehensive restructuring before any benefit emerges. The deployment operations leads can be named in a quarter; the quarterly review forum can be established in a month; the behavioral-monitoring function can be designated in weeks. The CAIO question, if it leads to appointment, takes longer — but the underlying ownership architecture should be in place before the appointment, not as a result of it.

Forward references

This chapter is the synthesis of the operational layer. The previous chapters named what to do — evaluation infrastructure, incident response runbooks, procurement contracts. This chapter names who specifically owns each of those artifacts. The operator who has read Modules 2 through 4 and applied the synthesis here has a coherent governance structure with explicit ownership at every level. Module 6, the regulatory posture chapter, treats the regulatory frameworks against which all of this operates.

Module 6: Regulatory Posture as a Living Operation

The how. What regimes apply, what they require, how to comply when the doctrine is still moving — and what to do when the deadline arrives and you are not ready.

The recommendation, stated first

Operating ahead of regulatory frameworks is no longer optional — the frameworks have arrived. Operating in compliance with them requires architectural and governance choices that most enterprise AI deployments have not made. The recommendation is to design every tier-3 and above deployment against the doctrine that is *plausibly evolving* rather than the rules that exist today, because the eighteen-month doctrine-evolution horizon is shorter than most procurement cycles and the cost of architectural reconstruction after a deadline is dramatically higher than the cost of building against the stricter interpretation from the start.

While the formal anchor for basic transparency remains August 2026, the EU's May 2026 political agreement on the AI Omnibus has officially deferred high-risk system obligations to December 2027 (for standalone systems) and August 2028 (for embedded systems). However, watermarking rules hit shortly after in December 2026, and the extraterritorial penalty structure remains aggressively higher than GDPR—topping out at 7% of global turnover.

In parallel, GDPR Article 22's *meaningful human involvement* doctrine continues to tighten through enforcement actions and member state DPA guidance. The CJEU SCHUFA judgment extended Article 22 to probability values where action taken on them significantly affects the data subject. The Dutch DPA initiated public consultation on operationalizing meaningful human intervention in March 2025. The doctrine is moving in one direction — toward broader scope and stricter interpretation. Operators planning around current minimum requirements are planning against a target that will not stay where it is.

Section 1: Regulatory classification

The first regulatory question for any agentic AI deployment is *which regimes apply to it.* The answer is rarely simple, and the cost of getting it wrong is asymmetric — over-classifying produces unnecessary compliance overhead; under-classifying produces material legal exposure that surfaces only when an incident or audit forces examination.

The classification has three independent dimensions that must be resolved separately.

The EU AI Act dimension. Does the deployment fall under the EU AI Act, and if so, at what risk level. The Act applies to providers and deployers of AI systems in the EU, plus extraterritorially to non-EU organizations whose AI outputs are used in the EU. The risk levels are: prohibited (limited applications, mostly inapplicable to enterprise contexts), high-risk (Annex III categories including employment, credit decisions, education, law enforcement, critical infrastructure, healthcare-adjacent applications), limited-risk (transparency obligations only — chatbot disclosure, deepfake labeling), minimal-risk (no specific obligations). A 2023 appliedAI study of 106 enterprise AI systems found 18% clearly high-risk and 40% with ambiguous classification — the ambiguous category is where most procurement-time misclassification happens.

The deployer-vs-provider question. Most enterprise operators are *deployers* under the AI Act, not providers. The distinction is structural and operationally critical. Providers develop AI systems or have them developed and place them on the EU market or put them into service internally under their own name or trademark; their obligations are extensive (Article 16). Deployers use AI systems in the course of professional activity; their obligations are narrower (Article 26) but still substantial — implementing technical and organizational measures to use the system per instructions, ensuring human oversight, monitoring operation against intended purpose, suspending use if there is reason to believe the system is non-compliant or poses risk, keeping logs for at least six months, informing affected workers and their representatives, and — if operating in banking, insurance, or public services — conducting fundamental rights impact assessments where applicable. Finally, they must cooperate with competent authorities. Crucially, under Article 25, deployers who substantially modify a high-risk system inherit the legal burdens of a Provider. The chapter recommends that every operator running an agentic AI deployment formally document its role classification and the specific obligations it has assumed.

The GDPR dimension. Independent of the AI Act, GDPR continues to apply to any deployment processing personal data. Article 22's automated-decision-making restrictions apply in parallel to AI Act transparency requirements; both regimes apply concurrently to AI systems processing personal data. The CJEU SCHUFA ruling extended Article 22 to probability values where an action taken on the value affects someone significantly. Implication: most enterprise AISDR systems that score prospects are within Article 22 scope when those scores drive material outreach decisions, even though the system is not making the literal "decision" itself.

The third dimension that matters in practice but does not cleanly classify is sectoral. Financial services have FINRA, the Digital Operational Resilience Act in the EU, and sector-specific privacy rules. Healthcare has HIPAA in the US and parallel frameworks elsewhere. Employment has NYC Local Law 144 on bias audits, Illinois HB 3773 on AI in hiring, and EEOC guidance. Each sectoral regime adds obligations that may overlap with or extend the AI Act and GDPR baseline. The chapter recommends that classification include a sectoral overlay specific to the deployment's use case.

The Monday action. Build a regulatory classification matrix for every agentic AI deployment in the portfolio. Each row is a deployment; the columns are EU AI Act applicability and risk level, deployer-vs-provider role, GDPR Article 22 applicability, applicable US federal frameworks, applicable state frameworks (Colorado AI Act if Colorado consumers in scope, California TFAIA if frontier model in scope, etc.), and applicable sectoral regimes. The matrix is a living document; review quarterly. The matrix's primary function is preventing the discover-the-regime-during-an-incident failure mode, which is the modal source of regulatory crisis.

Section 2: Documentation and technical evidence

Every applicable regime is fundamentally a documentation regime. The substantive obligations matter, but compliance demonstration requires evidence — and the evidence has to exist before the inquiry, not be constructed afterward.

The documentation requirements concentrate at the EU AI Act high-risk level. Article 11 plus Annex IV specify the technical documentation: general description of the AI system and its intended purpose, description of system components and processes, detailed information about training data and methodologies, description of system monitoring and human oversight, performance metrics, risk assessments, conformity assessment evidence, post-market monitoring plan. *As of April 2026*, the Commission has not finalized its standardization deliverables for high-risk system documentation; operators should monitor European standardization organizations for emerging standards while building documentation against current Annex IV requirements.

Three observations about the documentation requirements that affect how operators should build them.

The documentation requirements exceed traditional software documentation. The Annex IV list captures things most enterprise software organizations do not document — training data composition with sufficient detail for bias analysis, the specific risk assessments performed during development, the specific testing methodologies used, the mitigation measures for identified risks. Operators who treat AI documentation as an extension of standard software documentation will find the AI documentation deficient against AI Act requirements.

The documentation must be reproducible and current. The Annex IV requirement is not satisfied by point-in-time snapshots; the documentation must reflect the current state of the system, including changes since initial deployment. This is harder than it sounds when the system is updated frequently. The chapter's recommendation: treat the documentation as code, with version control, review cycles, and an explicit owner. Static documents stored in shared drives age into uselessness within months.

The documentation belongs in the procurement contract. This is where Module 4's discipline closes the loop. When the customer is the deployer and the vendor is the provider, the technical documentation that satisfies the deployer's obligations is generated by the provider. The procurement contract must require the provider to maintain Annex IV documentation, provide it to the deployer on request, and update it as the system evolves. Vendors who refuse this obligation are selling a system the deployer cannot operate compliantly.

The audit-trail requirement. Beyond Annex IV technical documentation, deployers must maintain logs sufficient to demonstrate compliance for at least six months under the AI Act, with longer retention for sector-specific regimes. The logging schemas from Module 3 should be designed against this requirement; the retention periods named there are the regulatory minima.

Section 3: Meaningful human involvement

The dominant doctrinal question across regimes is *what counts as meaningful human involvement* in automated decision-making. The doctrine has tightened steadily through enforcement and case law, and *as of April 2026* continues to tighten in a single direction: toward broader scope and stricter interpretation.

The legal anchor is GDPR Article 22(1): "The data subject shall have the right not to be subject to a decision based solely on automated processing, including profiling, which produces legal effects concerning him or her or similarly significantly affects him or her." Three operational sub-doctrines have emerged from enforcement.

The "solely" question. When does a decision count as "solely" automated such that Article 22 applies? The historical assumption was that any human in the loop satisfied the "not solely" requirement. That assumption has been dismantled. Courts and DPAs are now examining "the entire organizational environment where ADM is taking place, from the controller's organizational structure, to reporting lines and the effective training of staff." The reviewer must have the authority and competence to change the decision; the reviewer must have access to the information needed to evaluate; the reviewer must have volume allowing genuine consideration. All three apply concurrently.

The "decision" question. Per the CJEU's SCHUFA judgment, "decision" reaches further than literal final determinations. A probability value (credit score, risk score, prospect ranking) constitutes a decision under Article 22 if action taken on the value significantly affects the data subject. Implication for enterprise AISDR: a system that scores prospects and drives outreach decisions based on the scores is within Article 22 scope, even though the literal "decision" to contact the prospect may be made by a human salesperson based on the score.

The "significant effect" question. What counts as a legal or similarly significant effect? Legal effects include things like loan denials, insurance pricing, employment determinations, eligibility decisions for services. "Similarly significant" effects have been interpreted to include things that materially affect a person's choices, opportunities, or treatment. Targeted advertising based on inferences about behavior may qualify in some interpretations; differential pricing certainly does; algorithmic content moderation that affects platform access does.

The implication for tier-by-tier compliance posture is direct. *Tier 2 deployments* generally satisfy meaningful-human-involvement because the human is the author. *Tier 3 deployments* satisfy the doctrine only if the approval is meaningful by the three-criteria test — authority, information, volume. Most tier-3 deployments operating at high volume fail the volume criterion. *Tier 4 deployments* with sampled post-hoc review largely fail the doctrine under current interpretation, and operators relying on tier-4 architectures for regulated decisions should expect this position to be challenged. *Tier 5 deployments* can potentially satisfy meaningful human involvement through the boundary architecture itself — humans designed the boundary, humans approved it, humans maintain it — but this argument is not yet settled law and operators should design tier-5 boundary architectures against the stricter interpretation. *Tier 6 deployments* probably cannot satisfy meaningful human involvement under any plausible interpretation; operators considering tier 6 for regulated decisions should reconsider.

The active regulatory development worth tracking: the Dutch DPA's March 2025 public consultation on operationalizing meaningful human intervention. The fact that a major DPA initiated this consultation signals that the doctrine is being tightened, not relaxed. Operators should expect the bar to rise.

Section 4: Incident reporting

EU AI Act Article 73 establishes the most procedurally specific obligation in the entire Act. Providers of high-risk AI systems must report serious incidents to the market surveillance authorities of the Member States where the incident occurred. *As of April 2026*, the Commission has published draft guidance

(September 26, 2025) and a reporting template, with the final version expected to apply from August 2, 2026.

The reporting timelines are tight and asymmetric.

- **15 days** from awareness in the general case (e.g., fundamental rights or property damage).
- **10 days** from the moment a causal link is first suspected if the death of a person may have been caused.
- **2 days** from awareness in the event of a widespread infringement or serious, irreversible disruption of critical infrastructure.

Initial incomplete reports are acceptable, with supplemental information to follow. The deadlines run from the moment the provider establishes a causal link or the reasonable likelihood of one.

Three observations about Article 73 that affect operational readiness.

The scope is broad and includes indirect causation. The Commission's draft guidance clarifies that an indirect causal link between the AI system and the harm is sufficient to establish a duty to report. An incorrect medical analysis that causes harm only after a subsequent clinical decision is reportable. A loan denial based on flawed AI assessment is reportable. Disadvantageous treatment of qualified applicants through discriminatory selection is reportable. The Commission envisions broad scope; operators should not assume narrow interpretation.

The non-reporting penalty is separate from the underlying violation penalty. Failure to report under Article 73 carries administrative fines up to €15M or 3% of worldwide annual turnover — separately from any penalty for the underlying issue. The reporting obligation exists independently; an operator who responds to an incident competently but fails to report on time still faces the reporting penalty.

The investigation cooperation requirement is substantial. Following report, the provider must without delay perform necessary investigations including risk assessment and corrective action. The Commission's draft guidance specifies a prohibition on altering the AI system during the investigation without informing authorities if changes could affect subsequent analysis. Operators should plan for the post-incident period as a period of constrained operational change, not just a period of remediation.

A specific forward-looking concern worth flagging. Article 73 as currently drafted assumes single-system, single-occurrence failures. However, as the Cooperative AI Foundation established in its February 2025 technical report, the cumulative-effect harms generated by interacting AI systems are inadequately covered by the current single-agent paradigm. Operators of multi-agent systems should track the evolution of Article 73 in coming years and plan for likely expansion to cover multi-system incidents.

The Module 3 incident response runbook is where Article 73 compliance becomes operational. The runbook must include the reporting obligation as a first-class action, with named owners for the reporting decision, defined templates for the report, and integration with the cadence governance from Module 1. An organization that has Module 3 incident response infrastructure plus Article 73-aligned reporting processes is operationally ready for the regime; an organization with neither is exposed.

In parallel with Article 73, sector-specific reporting obligations continue to apply. Healthcare AI under Medical Device Regulation. Critical infrastructure under NIS2. Financial services under DORA. Some sectors permit Article 73 obligations to be discharged through equivalent sector-specific reporting; the deployer must verify equivalence rather than assume it.

Section 5: Cross-jurisdictional posture

Most enterprise operators face multiple regulatory regimes simultaneously: EU AI Act, GDPR, US federal frameworks, US state frameworks, sectoral regimes. The frameworks overlap substantially, conflict occasionally, and operate on different timelines.

The chapter recommends a portfolio-level posture rather than serial compliance.

Anchor against the strictest applicable regime. Where regimes overlap on a specific obligation, comply with the strictest. This is over-compliance against weaker regimes by definition, but the marginal cost is typically low and the protection against framework shifts is high. The August 2, 2026 EU AI Act high-risk obligations — effective from late 2027 — are the strictest applicable for most regulated enterprise deployments; building against them satisfies most US state requirements as well.

Treat NIST AI Risk Management Framework as the cross-jurisdictional baseline. *As of April 2026*, NIST AI RMF 1.0 is widely referenced and AI RMF 2.0 is in development. The Colorado AI Act explicitly designates NIST AI RMF as an affirmative defense against AG enforcement. ISO/IEC 42001 plays a similar role internationally. Operators who have implemented one of these frameworks have a posture that defends across multiple regimes. Operators who have not should prioritize one as the foundation for cross-jurisdictional defense.

The US state landscape is a moving patchwork. *As of April 2026*, the dominant state laws affecting enterprise agentic AI:

- **Colorado AI Act (SB 189, signed May 14, 2026)**, effective January 1, 2027. Repeals and replaces SB 24-205 (which had been delayed to June 30, 2026). Pivots from SB 24-205's risk-management and algorithmic discrimination framework to a transparency-based approach targeting Automated Decision-Making Technology (ADMT). Mandates consumer notices, automated adverse-outcome disclosures, and pathways for consumers to seek human review of algorithmic determinations.
- **California Transparency in Frontier AI Act (SB 53)**, effective January 1, 2026. Risk frameworks, critical safety incident reporting, specific safeguards for frontier developers.
- **California SB 942 (AI Transparency Act)**, watermarking requirements, delayed to August 2, 2026.
- **California sector-specific laws**: AB 489 (healthcare AI deceptive terms), SB 243 (companion chatbots), AB 325 (algorithmic price fixing).

- **Texas Responsible AI Governance Act**, effective January 1, 2026. Includes regulatory sandbox managed by Texas DIR.
- **Illinois HB 3773**, AI in employment disclosure and consent, effective February 2026.
- **NYC Local Law 144**, bias audits for automated employment decision tools. Continues active enforcement.

The federal preemption uncertainty. *As of April 2026*, Executive Order 14365 (December 11, 2025) attempts to establish federal preemption of state AI laws through executive action, including establishment of an AI Litigation Task Force and direction to federal agencies to challenge state AI laws on commerce-clause and preemption grounds. The White House National Policy Framework (March 20, 2026) recommends Congress enact preemption legislation. Critically, executive orders cannot independently preempt state law without congressional action; Congress has repeatedly declined to enact preemption (Cruz amendment defeated in OBBBA, July 2025; excluded from FY26 NDAA).

The chapter takes a position on the operational implication. **Prudent operators continue complying with state AI laws until courts rule otherwise.** The cost of compliance is recoverable through normal operations; the cost of non-compliance during a legal challenge is not recoverable, and the reputational damage of being on the wrong side of an enforcement action during pendency is real. Operators who bet on preemption are taking material risk that the preemption will resolve their way; operators who comply are taking the more defensible posture regardless of how preemption resolves.

The carve-outs from preemption ambitions under EO 14365 are worth noting: child safety in AI contexts, AI compute and data center infrastructure, state government procurement and use of AI. Operators in these areas should assume continued state enforcement authority.

The geopolitical-data-jurisdiction layer treated in Module 4 has regulatory implications here. Where data is processed, under which jurisdiction the vendor operates, what happens if the jurisdiction changes — these are not just procurement questions; they determine which regulatory regimes apply to specific data flows. Operators with cross-border deployments should map the data-jurisdiction layer explicitly and consult specialist counsel for high-stakes deployments.

Section 6: Affirmative defenses and risk-management frameworks

Most regulatory regimes permit affirmative defenses for operators who can demonstrate good-faith compliance with recognized risk-management frameworks. This is the primary mechanism by which operators convert compliance work into legal protection.

The dominant frameworks *as of April 2026*:

NIST AI Risk Management Framework (AI RMF). Voluntary US framework, but designated as affirmative defense in Colorado AI Act and influential federally. Builds on the NIST Cybersecurity Framework approach: govern, map, measure, manage. AI RMF 2.0 is in development; AI RMF 1.0 remains the operational anchor.

ISO/IEC 42001:2023. International standard for AI management systems. Plays a similar role to NIST AI RMF internationally and increasingly in the US for organizations preferring the ISO family. Operators with existing ISO 27001 information security management systems often find ISO 42001 a natural extension.

EU AI Act conformity assessment. For high-risk systems, conformity assessment is the formal compliance pathway. *As of April 2026*, the assessment infrastructure is still being built out — notified bodies are being designated, harmonized standards are being finalized, the assessment timelines for complex systems are 6-12 months. Operators planning for August 2, 2026 high-risk obligations — now deferred to December 2027 for standalone systems and August 2028 for embedded systems — should begin the conformity assessment process well in advance given that assessment timelines run 6-12 months for complex systems.

Sectoral frameworks. Sector-specific regimes have their own risk-management expectations: NIST CSF and SP 800-53 for federal contexts; FFIEC and OCC for financial services; HIPAA Security Rule and HHS guidance for healthcare; FERPA for education.

The chapter's recommendation: pick one risk-management framework as the organization's primary anchor (NIST AI RMF or ISO 42001 for most operators), implement it deliberately rather than performatively, and document the implementation in a form that supports affirmative-defense argument. Performative implementation — naming the framework in a policy without actually operating against it — produces the worst regulatory posture: the documentation is on file, but the substantive defense is unavailable when an inquiry surfaces the gap between policy and operation.

The chapter's specific guidance: the Module 1 cadence governance, the Module 2 evaluation harness, the Module 3 incident response infrastructure, the Module 4 procurement posture, and the Module 5 organizational design together constitute the operational implementation of NIST AI RMF or ISO 42001 for most enterprise deployments. Operators who have built the infrastructure across the previous chapters have largely implemented the framework; they must additionally document the implementation in framework-aligned form to claim the affirmative defense.

Section 7: Operating ahead of the frameworks

The chapter's organizing argument is that operators should design against doctrine that is *plausibly evolving* rather than against rules that exist today. This section makes the argument concrete.

The directional pressures *as of April 2026*:

Article 22 doctrine is tightening. The CJEU SCHUFA extension to probability values, the Dutch DPA consultation on operationalizing meaningful human intervention, the Article 73 incident reporting framework that captures indirect causation — all point in the same direction. Meaningful human involvement requirements are becoming stricter, not looser.

EU AI Act enforcement is converging on consistent interpretation across member states. The AI Office is providing centralized guidance; member state DPAs are coordinating. Operators who satisfy the strictest member state interpretation are largely safe across the EU.

US federal action remains uncertain. EO 14365, the National Policy Framework, and ongoing congressional inaction create uncertainty but not relief. State laws continue to enter force. The patchwork is growing, not shrinking.

Sectoral frameworks are extending their AI-specific obligations. FINRA, EEOC, HHS, and parallel international regulators are all developing more specific AI guidance. Operators in regulated sectors should expect sector-specific obligations to expand throughout the chapter's shelf life.

Multi-agent and self-modifying system regulation is forming. The current regulatory frameworks were largely designed against single-system AI. The Cooperative AI Foundation critique of Article 73 multi-agent inadequacy is one example; expect similar gaps to be addressed through guidance and amendment over coming years.

The anticipatory move is to build tier-3 discipline at tier 4, tier-4 discipline at tier 5, tier-5 discipline at tier 6 — and to build current-doctrine compliance against the stricter plausible interpretation rather than the current minimum. Concretely:

- *At tier 3*, build reviewer-effectiveness measurement as if the meaningful-human-involvement bar will rise (it is).
- *At tier 4*, build sampled-review architectures as if regulators will examine whether the sampling produces meaningful oversight (they are starting to).
- *At tier 5*, build boundary architectures with the duty-of-care frame in mind (it is becoming the dominant theory of liability).
- *At tier 6*, build policy-versioning and drift-detection infrastructure as if regulatory frameworks will catch up to the technology (they are).

The cost of this posture is real. Building against stricter interpretations is more expensive than building against current minimums. The cost of not adopting this posture is also real and arrives later: architectural reconstruction after a deadline or an enforcement action is dramatically more expensive than building against the stricter interpretation from the start.

Emergency triage: if you are reading this and are not compliant

Some readers will encounter this chapter and discover their deployments are not compliant with obligations that are already in force or approaching. The moment does not matter; the response does. The chapter includes this section deliberately because regulators tend to penalize willful non-compliance more harshly than acknowledged-and-remediating non-compliance.

Immediate actions if you discover non-compliance:

Document the discovery — when, by whom, what was discovered, what the gap is. The documentation supports good-faith remediation argument if subsequent enforcement action occurs.

Triage the deployments. Which are highest risk by tier (tier 4+), by use case (regulated decisions about individuals), and by jurisdiction (EU operations, Colorado consumers, sector-specific regulated activities). The highest-risk combinations get attention first.

Implement the documentation requirements first. Annex IV technical documentation, deployer logs, processing records — these are the artifacts regulators will request first. Even imperfect documentation is better than absence.

Suspend deployments that cannot be brought to defensible posture quickly. Suspended-pending-remediation is a defensible posture; operating-non-compliantly is not. The decision is hard because suspension has business cost; the decision is correct when the alternative is escalating regulatory exposure.

Engage counsel immediately. The regulatory regimes covered in this chapter are jurisdiction-dependent and use-case-dependent in ways no chapter can fully resolve. Counsel-led remediation is dramatically more defensible than self-led remediation, particularly when reporting obligations attach.

For Article 73 specifically: if a serious incident has occurred and was not reported within the deadline, report immediately with explanation of the delay. Late reporting is better than non-reporting; the underlying penalty is separate from the reporting penalty, and demonstrated good-faith remediation typically reduces both.

Forward references

The cross-references close the loop on the modular structure. **Module 1** established the autonomy gradient and tier vocabulary that this chapter has applied throughout. **Module 2** specified the evaluation infrastructure that supports regulatory defense — the integrity dashboard, the performance harness, the drift detection, the cadence layer. **Module 3** specified the logging schemas that satisfy documentation requirements and the incident response runbooks that operationalize Article 73 compliance. **Module 4** specified the procurement contracts that obligate vendors to support customer regulatory compliance — the data rights, audit rights, post-deployment monitoring obligations, and incident-cooperation requirements. **Module 5** specifies the organizational design that names the humans responsible for each regulatory function — the deployer-obligation owner, the meaningful-human-involvement reviewer, the Article 73 reporting decision-maker, the affirmative-defense documentation owner.

A reader who has built the infrastructure across the previous modules has largely implemented the regulatory posture this chapter describes. The remaining work is documentation, framework alignment, and named ownership — the operational completion of the architectural commitments.

The frameworks will continue to evolve. The chapter is written to age at the level of doctrine; specific provisions will be superseded. The recommendation throughout — design against plausibly evolving doctrine rather than current minimum — is what makes the architectural choices durable across the regulatory shifts that *will* occur in the coming years.

Closing: The Discipline Outlives the Frameworks

The book's argument in mature form, the trilogy's accountability questions resolved, and what comes next.

What the book has earned the right to say

The introduction asked the reader to consider whether the deployment they were claiming to operate was the deployment that was actually running. Six modules of operational reference and four essays of accountability argument later, the answer is now structural rather than rhetorical. The deployment you are claiming to operate is *almost certainly not* the deployment that is actually running, in most enterprises, most of the time. The drift has happened. The gap exists. What changes after this book is not whether the gap is real but whether the operator can detect it, measure it, and close it.

The book's central argument can now be stated in mature form. *Agentic AI is not governed by policy; it is governed by cadence.* The artifacts most organizations produce — risk assessments, ethics charters, model cards, governance frameworks — describe a deployment at a moment in time. They do not, on their own, hold a deployment at any operational position. What holds a deployment is the discipline of weekly behavioral metric review, monthly trend analysis, quarterly governance review, semi-annual portfolio review — each with named owners, defined authority, and the operational infrastructure that makes the cadence enforceable rather than theatrical. The book has named the metrics, the owners, the artifacts, and the cadences that constitute this discipline at each tier of the autonomy gradient. An organization that has implemented the discipline has the means to detect drift, contain it when it occurs, and govern its agentic AI deployments at the position the architecture decision record actually describes. An organization that has the policy without the cadence has documented a deployment it is not actually governing.

This argument was anticipatory when the introduction made it. After six operational chapters, it is supported. The reader who has worked through the book has the discipline. What remains is to apply it.

The trilogy resolved

Part I named four accountability problems that the operational chapters have now answered. The resolution is worth making explicit because the connection between Part I's rhetorical argument and Parts II and III's operational discipline is the book's load-bearing structural move.

The overture argued that autonomous AI sales agents are the third thing operators should build, not the first, and that the order in which most organizations are pursuing agentic AI reveals where their accountability question has been quietly deferred. The operational discipline answers the deferral. *Tier-claim verification* (Module 4's procurement contribution) reveals when vendor-marketed deployment tier exceeds operational tier. *Diagnostic metrics* (Module 2's evaluation contribution) make the operational tier measurable rather than asserted. *Article 22 doctrine application* (Module 6's regulatory contribution) names what the meaningful-human-involvement bar requires at each tier. An operator who has built the discipline does not have to defer the accountability question to a later moment of crisis; the question is being answered weekly, in the data the operator can compute against their own production logs.

The targeting essay argued that the accountability question begins before the system acts — at the choice of who the system is built to act on, and what data justifies that choice. The operational discipline answers this through procurement (Module 4) and regulatory posture (Module 6). The procurement contract obligates vendors to provide the provenance documentation and audit rights that make targeting decisions defensible against regulatory inquiry. The regulatory posture chapter specifies the data-protection regimes (GDPR, Article 22, Colorado AI Act, the data broker framework) that govern what can be collected, what can be inferred, and what targeting decisions require justification. Operators who have built both have the upstream accountability the essay argued was missing.

The utterance essay argued that copilots structurally relocate accountability from the system to the human in the loop, and that the human's seat absorbs liability the architecture is designed to produce but not to bear. The operational discipline answers this through evaluation infrastructure (Module 2) and incident response (Module 3). The evaluation infrastructure measures whether the human review is meaningful — through reviewer latency, through edit ratio, through the diagnostic metrics that distinguish substantive review from rubber-stamping. The incident response chapter specifies how to detect compliance drift before it becomes a regulatory matter, and how to allocate accountability between system and operator when the drift produces harm. Operators who have built both can demonstrate the meaningful human involvement that Article 22 doctrine increasingly demands.

The outcome essay argued that self-improving AI breaks the assumption of stable agent identity that classical principal-agent theory requires, and that four parties — vendor, buyer, prospect, model — can each plausibly disclaim accountability for outcomes a system produces after it has rewritten itself. The operational discipline answers this through Tier 6 governance (Module 1) and post-market monitoring contracts (Module 4). The Tier 6 chapter specifies what governance must look like for systems that modify their own policy: policy versioning, drift detection, the boundary architecture that constrains what self-modification can do. The procurement chapter specifies the contractual obligations that obligate vendors to provide the audit rights and reporting that make accountability allocable rather than diffuse. Operators who have built both have the temporal accountability the essay argued was structurally missing.

The trilogy named the problems. The operational chapters specified the discipline. The closing of the loop is what makes the book a coherent argument rather than a collection of essays plus a reference work. After the operational chapters have done their work, the trilogy's questions are answerable — not in the abstract, but in the specific artifacts the book has named.

What comes next

Three direction-of-evolution claims about where the field is going, stated narrowly because narrow forward-looking claims age better than broad ones. The reader will encounter evolution after the book is published; what follows is the frame for navigating it.

The meaningful-human-involvement doctrine will continue tightening. *As of April 2026*, the trajectory is unmistakable. The CJEU's SCHUFA judgment extended Article 22 to probability values producing significant action. The Dutch DPA initiated public consultation in March 2025 on operationalizing meaningful human intervention. The Future of Privacy Forum's analysis of GDPR enforcement found that data protection authorities are now examining "the entire organizational environment where automated decision-making is taking place, from the controller's organizational structure, to reporting lines and the effective training of staff." The bar for meaningful human involvement is rising, not falling. Operators planning around current minimum requirements are planning against a target that will not stay where it is. The book's diagnostic metrics — reviewer latency, edit ratio, propagation latency — give operators the means to demonstrate meaningful involvement rather than merely assert it. Operators who can demonstrate will have a defensible posture;

operators who can only assert will face the doctrine's tightening as risk they cannot manage.

The deployment-vs-production gap will not close quickly. *As of April 2026*, multiple converging surveys document the same pattern. WRITER's 2026 enterprise AI adoption survey reports that 97 percent of executives deployed AI agents in the past year, but only 12 percent of agent initiatives reach production at scale. Stanford's enterprise AI playbook of 51 successful deployments found that two-thirds had significant failed attempts before achieving value. Industry analysis attributes 89 percent of scaling failures to integration complexity, output quality at volume, monitoring gaps, organizational ownership ambiguity, and insufficient domain training data — every one of which is an operational discipline gap rather than a model capability gap. The deployment-vs-production gap is structural; closing it requires the kind of work this book teaches. The gap will likely persist through 2027 and 2028, which means the operational discipline this book describes will remain a competitive advantage rather than a baseline. Operators who build it will ship; operators who do not will continue to add to the pilot inventory without producing operational AI.

The doctrine-vs-instance discipline will become more important, not less. As more enforcement actions land, more case law develops, more guidance documents publish, more state laws activate, and more federal preemption attempts proceed, the volume of point-in-time material will overwhelm any operator's ability to track everything. The Mobley v. Workday opt-in deadline closed March 7, 2026, and the case proceeds to merits with a defined nationwide collective; a parallel case against Eightfold AI was filed January 20, 2026; the EU AI Act's August 2026 deadline is approaching with the Digital Omnibus possibly delaying high-risk obligations to December 2027 or August 2028. Federal preemption legislation is in committee with uncertain prospects — the TRUMP AMERICA AI Act and the GUARDRAILS Act represent opposing positions, neither likely to pass quickly. The point-in-time picture will look different in six months than it does today; it will look different again six months after that. The operational answer is to track doctrine rather than to track instances. The Air Canada principle, the meaningful-human-involvement standard, the duty-of-care frame for boundary architecture, the agent-doctrine extension to vendor-supplied AI — these are durable. Specific provisions and specific cases will move; the doctrine they embody is what operators must internalize and apply.

These three directions are stated narrowly on purpose. The book is not predicting specific outcomes. It is naming the trajectory, naming the operational discipline that responds to the trajectory, and trusting the reader to track the specifics through their own current research at the time of reading. The doctrine-vs-instance discipline applied to forward-looking writing produces the same posture the book has taken throughout: anchor at the level of doctrine, mark instances as point-in-time, build operational infrastructure that survives shifts at the instance level.

What the book asks of the reader

A reference work is judged by what readers do after they finish it. This book has been written to be applied, not merely understood. Three things the reader who has worked through the book is now equipped to do.

Detect. Run the queries against your own production logs that the operational chapters specified. The behavioral metrics — reviewer latency at Tier 3, propagation latency at Tier 4, exception-handling time at Tier 5 — are computable from data your organization already produces. The first detection is usually the most consequential one, because it reveals whether the deployments your organization claims to operate match the deployments that are actually running. Most operators discover the answer is no.

Govern. Implement the cadence at the right depth for your tier. Weekly behavioral review at the operational team level. Monthly trend analysis at the architecture decision record owner level. Quarterly governance review at the executive level. Semi-annual portfolio review at the senior governance forum. The cadences are scaled to deployment risk — proportionality is the discipline. An organization operating tier-3 deployments at a small number of low-risk use cases does not need the full cadence; an organization operating tier-5 deployments at scale across regulated domains cannot operate without it.

Defend. Build the documentation, the affirmative defenses, and the procurement contracts that demonstrate good-faith operation when scrutiny arrives. Annex IV technical documentation against the EU AI Act. NIST AI RMF or ISO 42001 alignment as affirmative defense. Contractual obligations that allocate vendor and customer responsibilities defensibly. The discipline is not just internal — it is also the substrate that makes external accountability possible when an inquiry, an audit, or a regulatory action requires the organization to demonstrate that it operated responsibly.

These three actions — detect, govern, defend — are the operational summary of the book's argument. An organization that has implemented all three has built the discipline. An organization that has implemented none of them is exposed in ways that the regulatory frameworks, the case law, and the failure cases now in motion will eventually surface.

A final framing

The book has named a discipline. The discipline is not the autonomy gradient itself — that is the analytic vocabulary. The discipline is the operational practice the gradient enables: knowing what tier each deployment operates at, measuring whether the operation matches the claim, holding deployments at their declared positions through cadence rather than through policy. The framework is the language; the discipline is the work.

The frameworks will change. The autonomy gradient resolution in this book is one of several being proposed in 2025 and 2026; future authors will propose others. The specific provisions of the EU AI Act, the doctrine of GDPR Article 22, the federal preemption status, the case law of agent-vs-tool — all of these will look different in 2027, in 2028, in 2030. What survives is the underlying operational practice: the cadence, the diagnostic metrics, the integrity dashboards, the post-market monitoring obligations, the cross-functional ownership architectures. These are not products of the book's specific framework. They are the operational answer to a structural problem in agentic AI governance, and they apply regardless of which framework the operator uses to describe their deployments.

This is why the book is written to outlive its specific frameworks. The reader who comes back to the book in 2028 will find provisions that have been superseded, cases that have been decided, frameworks that have been extended or replaced. The reader will also find the operational discipline still applies — because the underlying problem the discipline addresses has not gone away. Agentic AI deployments will continue to drift from their declared positions. Policy without cadence will continue to fail. The mathematical relationship between deployment volume and meaningful human review will continue to produce the rubber-stamping that the architecture was designed to prevent. The discipline that addresses these problems is durable in a way that specific provisions are not.

The agent your enterprise deploys tomorrow will not be the agent your enterprise is operating six months from now. The book's contribution is the discipline that makes the gap visible — and closable.

That is the work. The book has named it. The reader is now equipped to do it.

Glossary

Terms defined as the book uses them. The glossary is for reference rather than reading; entries are alphabetized and intended to be consulted in any order. Terms that the book introduces or develops distinctively are marked (this book); other entries define how the book uses terms drawn from the regulatory frameworks, case law, or broader field.

Accountability object (this book) — The specific artifact, decision, or process for which a particular human is responsible at a given autonomy tier. The accountability object differs by tier: the artifact at Tier 2, the approval at Tier 3, the action at Tier 4, the boundary at Tier 5, the adaptation process at Tier 6. Identifying the accountability object is the first step in establishing meaningful governance at that tier.

Agentic AI — AI systems capable of taking actions in the world, not just producing outputs for humans to consider. Used in this book to denote systems operating at Tier 2 and above on the autonomy gradient, where the system's output enters operational workflows rather than ending at the human reviewer.

Agentic tier (this book) — A specific position on the autonomy gradient, defined by how the system acts in the world, who is doing the runtime safety work, and what the accountability object is. Six tiers: Tier 1 (Suggestion), Tier 2 (Draft-and-Edit), Tier 3 (Approve-and-Send), Tier 4 (Supervised Execution), Tier 5 (Bounded Autonomy), Tier 6 (Self-Modifying Autonomy). Replaces the binary "human in the loop / autonomous" framing and the three-level "assistive / supervised / autonomous" framing common in the field.

AI Act (EU) — Regulation (EU) 2024/1689, the European Union's comprehensive framework regulating artificial intelligence. Phased into application: prohibited practices effective February 2025, GPAI obligations August 2025, high-risk system obligations August 2026, regulated-product systems August 2027. Penalties up to €35M or 7% of worldwide turnover for prohibited practices.

AISDR — Autonomous AI sales development representative. AI system that researches prospects, drafts outbound communications, and (at higher tiers) sends them. Used in this book as the primary case study throughout Part I because AISDR exhibits the full autonomy gradient and has the richest set of public failure cases in 2026.

Annex III (EU AI Act) — The list of AI system categories classified as high-risk under the EU AI Act, including employment-related systems, credit and insurance decisions, education access, law enforcement uses, critical infrastructure operation, and healthcare-adjacent applications.

Annex IV (EU AI Act) — The technical documentation requirements for high-risk AI systems. Specifies what providers must document: system description, components and processes, training data and methodologies, monitoring and human oversight, performance metrics, risk assessments, conformity evidence, post-market monitoring plans.

Approval inflation (this book) — The pattern at Tier 3 where reviewer latency compresses toward zero as deployment volume increases, producing approvals that have the architectural form of meaningful review without the operational substance. The mathematical relationship between deployment volume and human review capacity makes approval inflation predictable when capacity is not scaled to volume.

Article 22 (GDPR) — The provision of the General Data Protection Regulation establishing data subjects' right not to be subject to decisions based solely on automated processing that produce legal effects or similarly significant effects. The CJEU's SCHUFA judgment (Case C-634/21) extended Article 22 to probability values producing significant action.

Article 73 (EU AI Act) — The provision establishing serious-incident reporting obligations for providers of high-risk AI systems. Reporting timelines: 15 days general, 10 days if death may have been caused, 2 days for widespread infringement or critical infrastructure disruption. Indirect causal links suffice to establish reportability.

Authoritative-record propagation latency (this book) — The Tier 4 tier-integrity metric. Measures the time between a fact changing in an authoritative source-of-truth system and every autonomous agent surface reflecting the change. Increasing latency is the drift signal toward un-governable execution. Also called the *dialer check*.

Behavioral drift (this book) — The default trajectory of any agentic AI deployment, in which user behavior over time produces operational migration to a higher autonomy tier than the architecture committed to. The deployment that begins at Tier 3 (approve-and-send) drifts toward Tier 4 (rubber-stamping) without architectural change. Drift is the default, not the exception, and is detected through tier-integrity metrics rather than through architectural review.

Boundary architecture (this book) — The Tier 5 system design pattern in which the agent operates autonomously within declared, operationally-enforced constraints. The boundaries are documented, version-controlled, and machine-enforced — not described in design documents. The exception-handling path is built, staffed, and measured. At Tier 5, the boundary is the accountability object.

Boundary-exception handling time (this book) — The Tier 5 tier-integrity metric. Measures the median and 95th-percentile time between an exception triggering and meaningful handling. Degradation of this metric signals silent migration to either Tier 4 (exceptions stop being handled meaningfully) or Tier 6 (the boundary is being loosened informally to reduce exception volume).

Cadence governance (this book) — The discipline of maintaining agentic AI deployments at their declared autonomy tiers through scheduled measurement and review rather than through policy documents. Four cadence levels: weekly behavioral metric review, monthly trend analysis, quarterly governance review, semi-annual portfolio review. The book's central operational claim is that governance is cadence, not policy.

CAIO (Chief AI Officer) — Executive role responsible for AI strategy and operational governance. As of 2026, approximately 76% of large organizations have a CAIO, up from 26% the prior year. The book takes the position that the role is appropriate for large enterprises with cross-functional integration gaps, possibly transitional even at large enterprises, and not appropriate for organizations under approximately $1B revenue.

Conformity assessment (EU AI Act) — The formal compliance pathway for high-risk AI systems under the EU AI Act. As of April 2026, the assessment infrastructure is still being built: notified bodies are being designated, harmonized standards finalized. Assessment timelines run 6-12 months for complex systems.

Cooperative AI Foundation — Research organization whose January 2026 critique observed that EU AI Act Article 73 as drafted assumes single-system, single-occurrence failures, leaving multi-agent systems and cumulative-effect harms inadequately covered. The book treats this critique as a forward-looking concern operators of multi-agent systems should track.

Deployer (EU AI Act) — Under the EU AI Act, an entity that uses an AI system in the course of professional activity, distinct from a provider that develops the system. Most enterprise operators are deployers. Article 26 specifies deployer obligations: implementing technical and organizational measures, ensuring human oversight, monitoring operation, suspending use when non-compliance is suspected, keeping logs for six months, informing affected workers, conducting fundamental rights impact assessments where applicable.

Dialer check — See *authoritative-record propagation latency*.

Diff logging — The Tier 2 logging schema in which the difference between model-generated content and human-edited final content is captured at every send event. Diff logs enable the edit ratio metric and reconstruct the editing trajectory if a downstream incident requires it.

Doctrine-vs-instance discipline (this book) — The editorial and analytical practice of anchoring durable claims at the level of doctrine while marking specific provisions, cases, or guidance documents as point-in-time markers that may shift. Applied throughout the book: italicized "as of [date]" references mark time-limited claims; the underlying doctrinal arguments are intended to age well across regulatory shifts.

Edit ratio (this book) — The Tier 2 tier-integrity metric. Measures the fraction of model-generated content replaced or modified by human editing in the final sent artifact. A low edit ratio signals drift toward Tier 3 (model-produces-and-human-ratifies). Reasonable starting threshold: deployment is functioning as Tier 3 below median 30% edit ratio; the Tier 2 architectural claim is no longer supportable below 15%.

EO 14365 — Executive Order issued December 11, 2025, "Ensuring a National Policy Framework for Artificial Intelligence." Establishes the AI Litigation Task Force to challenge state AI laws on commerce-clause and preemption grounds. Directs federal agencies to issue policy statements, evaluate state laws, and condition federal funds on state AI law posture. Implementation *as of April 2026* is largely stalled; congressional preemption legislation has not advanced.

FRIA (Fundamental Rights Impact Assessment) — Required of certain deployers of high-risk AI systems under the EU AI Act, integrated with DPIA where personal data is processed.

Hancock declaration ruling — Minnesota federal court, January 2025. Judge Laura Provinzino excluded an expert declaration from a Stanford professor after fabricated citations were discovered in his GPT-4o-assisted submission. Provinzino's ruling — particularly the language that experts must not "abdicate their independent judgment" in favor of AI-generated answers — is the cleanest judicial articulation of the copilot accountability problem. Part of the Mata v. Avianca lineage establishing human responsibility for AI-mediated work product.

Marketed-vs-operational tier gap (this book) — The discrepancy between the autonomy tier a vendor markets a deployment at and the tier the deployment actually operates at. Vendors of agentic AI commonly market deployments at Tier 5 or Tier 6 ("digital workers") while the operationally-defensible tier is Tier 3 or Tier 4. The book treats the gap as a material concern that procurement contracts should address through tier-claim verification clauses.

Mata v. Avianca, Inc. — Southern District of New York, 2023. Judge P. Kevin Castel sanctioned attorneys who submitted a legal brief containing fabricated case citations generated by ChatGPT, then submitted a second round of fabrications when challenged. The case established the doctrinal principle that humans using AI tools are responsible for the outputs they deliver, regardless of how those outputs were generated. The Mata principle informs the book's treatment of Tier 3 governance and connects to GDPR Article 22's meaningful-human-involvement doctrine.

Meaningful human involvement — The doctrinal standard, drawn from GDPR Article 22 enforcement and member state DPA guidance, for what counts as human oversight of automated decision-making. Three operational sub-criteria: the reviewer must have authority to change the decision, the information necessary to evaluate, and volume allowing genuine consideration. Doctrine is tightening through enforcement; the bar is rising.

MVG (Minimum Viable Governance) — Framework adapted from NIST AI RMF for organizations under approximately 500 employees that cannot sustain enterprise-scale governance infrastructure. Four pillars: Govern, Map, Measure, Manage. Anchored on a 1-Page AI Ethics Charter with explicit "red lines" and core commitments. Treated in Module 5 as the appropriate substitute for full cadence governance at startup scale.

NIST AI RMF (AI Risk Management Framework) — Voluntary US framework published by the National Institute of Standards and Technology. AI RMF 1.0 is the operational anchor as of April 2026; 2.0 is in development. Designated as affirmative defense in Colorado AI Act and influential federally. Builds on NIST Cybersecurity Framework structure: govern, map, measure, manage.

Operation AI Comply — FTC enforcement initiative launched September 2024, applying Section 5 of the FTC Act to AI marketing claims and AI-driven harms. Approximately a dozen cases through 2025 and into 2026, including DoNotPay, Ascend Ecom, Ecommerce Empire Builders, FBA Machine, Rytr (later reopened), IntelliVision, Workado, Click Profit, and Air AI.

Policy-drift detection latency (this book) — The Tier 6 tier-integrity metric. Measures the time between a self-modifying system updating its policy and the operator detecting the policy change. At Tier 6, the accountability object is the adaptation process itself; without policy-drift detection, the operator cannot exercise governance over what the system has become.

Post-market monitoring — EU AI Act Article 72 obligation for providers of high-risk AI systems to systematically monitor performance after deployment. The book treats post-market monitoring as a contractual obligation that procurement contracts should obligate vendors to support, including providing data necessary for the customer's own deployer-side compliance.

Provider (EU AI Act) — Under the EU AI Act, an entity that develops AI systems or has them developed and places them on the EU market under its own name or trademark. Provider obligations are extensive (Article 16). Distinct from deployer.

Reviewer latency (this book) — The Tier 3 tier-integrity metric. Measures the time between when an action is presented to a human reviewer and when the reviewer approves it. Distribution compression toward zero is the drift signal indicating approval inflation. Tracked per reviewer, per action type, in aggregate; alerts fire when the distribution falls below the meaningful-review threshold.

RLAIF (Reinforcement Learning from AI Feedback) — Training technique in which an AI labeler model provides the reward signal for a policy model, replacing human evaluators. Enables systems that improve themselves without humans in the loop. Inherits classical reward-hacking risks and adds new ones specific to AI-as-judge supervision (amplified bias, sycophancy, reward signal noise from labeler hallucination).

Rubber-stamping — The pattern at Tier 3 where reviewer behavior compresses to clicking approve without substantive review. Mathematically inevitable when deployment volume exceeds the human capacity allocated to meaningful review.

SCHUFA (CJEU Case C-634/21) — Court of Justice of the European Union judgment extending GDPR Article 22 to probability values where action taken on them significantly affects the data subject. Implication: most enterprise

AISDR systems that score prospects are within Article 22 scope when those scores drive material outreach decisions.

Section 5 (FTC Act) — Section 5 of the Federal Trade Commission Act, prohibiting unfair or deceptive acts or practices. Applied by the FTC to AI marketing claims and AI-driven harms under Operation AI Comply, demonstrating that existing consumer protection law applies to AI-driven outreach without waiting for AI-specific legislation.

Tier-claim verification (this book) — The procurement discipline of contractually requiring vendors to provide the data necessary for the customer to compute tier-integrity metrics on the deployment. Tier-claim verification clauses obligate vendors to support the customer's measurement of whether the deployment is operating at the marketed tier, with the marketed-vs-operational gap as a material breach if it exceeds threshold.

Tier-integrity metric (this book) — A single behavioral measurement for each autonomy tier that, if it drifts, signals the deployment has silently moved to a different tier. The catalog: Tier 2 edit ratio, Tier 3 reviewer latency, Tier 4 authoritative-record propagation latency, Tier 5 boundary-exception handling time, Tier 6 policy-drift detection latency. Tier-integrity metrics are computable from production logs and are the operational instrument by which behavioral drift becomes visible.

Sources and Citations

Sources organized by category. Where the book engages a source substantially, a brief annotation indicates how. Specific provisions, case citations, and survey numbers should be verified against current sources at the time of reading; the materials below were current as of April 2026.

Regulatory Frameworks and Laws

European Union

Regulation (EU) 2024/1689 (EU AI Act). The European Union's comprehensive framework for AI regulation. Phased application: prohibited practices February 2025, GPAI obligations August 2025, high-risk system obligations August 2026, regulated-product systems August 2027. The book engages Articles 11, 16, 22, 26, 27, 50, 72, 73, plus Annexes III and IV throughout the regulatory chapter (Module 6).

Article 73 Draft Guidance (European Commission, September 26, 2025). 36-page guidance document plus reporting template, covering serious-incident reporting timelines, indirect causation, and operator obligations. Public consultation closed November 7, 2025.

Digital Omnibus Proposal (European Commission, November 19, 2025). Proposed legislative package potentially delaying high-risk AI system obligations to long-stop dates of December 2027 and August 2028.

Regulation (EU) 2016/679 (GDPR). General Data Protection Regulation. The book engages Article 22 (automated decision-making) extensively in connection with the meaningful-human-involvement doctrine.

Dutch DPA Public Consultation on Meaningful Human Intervention (March 6, 2025). Regulator-led consultation on operationalizing Article 22's human-involvement requirement.

United States — Federal

Executive Order 14365 (December 11, 2025). "Ensuring a National Policy Framework for Artificial Intelligence." Establishes the AI Litigation Task Force, directs federal policy statements, and conditions BEAD federal funds on state AI law posture. 90 Fed. Reg. 58499.

White House National Policy Framework for AI (March 20, 2026). Recommends congressional preemption legislation. As of April 2026, no congressional action.

TRUMP AMERICA AI Act (introduced March 18, 2026). 291-page federal preemption legislation. In committee; uncertain prospects.

GUARDRAILS Act (introduced March 20, 2026). Legislation to repeal Executive Order 14365. In committee.

OMB M-24-10 (March 28, 2024). Federal CAIO mandate for U.S. agencies. Sets seniority requirements, reporting structures, and coordination obligations referenced as a public-sector benchmark in Module 5.

Section 5 of the FTC Act (15 U.S.C. § 45). Prohibits unfair or deceptive acts or practices. Applied by the FTC under Operation AI Comply to AI marketing claims and AI-driven harms.

TAKE IT DOWN Act. Provisions effective May 2026, addressing AI-generated synthetic media.

United States — State

Colorado AI Act (SB 189, signed May 14, 2026), effective January 1, 2027. Repeals and replaces SB 24-205 (which had been delayed to June 30, 2026).

Pivots from SB 24-205's risk-management and algorithmic discrimination framework to a transparency-based approach targeting Automated Decision-Making Technology (ADMT). Mandates consumer notices, automated adverse-outcome disclosures, and pathways for consumers to seek human review of algorithmic determinations.

California Transparency in Frontier AI Act (SB 53). Effective January 1, 2026. Risk frameworks, critical safety incident reporting, specific safeguards for frontier developers.

California AI Transparency Act (SB 942). Watermarking requirements; effective date August 2, 2026 (delayed from January 1, 2026).

California AB 489. Healthcare AI deceptive terms; prohibits AI from falsely claiming healthcare licenses.

California SB 243 (Companion Chatbots Act). Chatbot disclosures, safety protocols, minor protections.

California AB 325 (Preventing Algorithmic Price Fixing Act). Updates Cartwright Act to bar shared pricing algorithms.

California Delete Act and DROP (Data Removal and Opt-Out Platform). DROP launch January 1, 2026.

Texas Responsible AI Governance Act (RAIGA). Effective January 1, 2026. Includes AI regulatory sandbox managed by Texas DIR.

Illinois HB 3773. AI in employment disclosure and consent requirements; effective February 2026.

New York City Local Law 144. Bias audits for automated employment decision tools.

New York RAISE Act. Various 2026 provisions including synthetic-performer disclosure.

Voluntary Frameworks and Standards

NIST AI Risk Management Framework (AI RMF 1.0). National Institute of Standards and Technology, January 26, 2023. Designated as affirmative defense under Colorado AI Act and influential federally. AI RMF 2.0 in development as of April 2026.

ISO/IEC 42001:2023. International standard for AI management systems. International equivalent role to NIST AI RMF.

GSA Proposed AI Clause (GSAR 552.239-7001). Published March 6, 2026; comment period extended to April 30, 2026. Federal procurement template.

Case Law and Enforcement

Mobley v. Workday. N.D. Cal. Case No. 23-cv-00770-RFL. July 2024 ruling allowing claims to proceed on agent theory. May 16, 2025: Judge Rita Lin granted preliminary certification of a nationwide collective action under the Age Discrimination in Employment Act, with notice to potentially hundreds of millions of class members. July 29, 2025: Scope expanded to include applicants processed using HiredScore AI features. Opt-in deadline closed March 7, 2026; case proceeds to merits.

Anti-Eightfold AI Litigation (filed January 20, 2026). Parallel AI hiring discrimination case extending the agent-doctrine question to a different vendor.

Moffatt v. Air Canada (2024 BCCRT 149). British Columbia Civil Resolution Tribunal, February 14, 2024. Tribunal Member Christopher C. Rivers held airline responsible for AI-generated chatbot content; rejected the "separate legal entity" defense. $812.02 total in damages and fees. The Air Canada principle is the doctrinal anchor.

Mata v. Avianca, Inc. S.D.N.Y., 2023. Judge P. Kevin Castel sanctioned attorneys who submitted ChatGPT-generated brief with fabricated case citations. The lineage of Hancock declaration cases and AI hallucination accountability.

Hancock Declaration (Minnesota federal court, January 2025). Judge Laura Provinzino excluded an expert declaration from a Stanford professor after fabricated citations were discovered in his GPT-4o-assisted submission. Provinzino's "abdicate their independent judgment" language is the cleanest judicial articulation of the copilot accountability problem.

SCHUFA Holding AG v. OQ (CJEU Case C-634/21). Court of Justice of the European Union, December 7, 2023. Extended GDPR Article 22 to probability values producing significant action.

FTC Operation AI Comply Cases

Initial five cases (September 25, 2024): DoNotPay ($193,000 settlement), Ascend Ecom, Ecommerce Empire Builders, FBA Machine, Rytr.
IntelliVision (December 3, 2024). Settled deceptive marketing claims about facial recognition system bias.
Workado (April 2025). AI content detection deceptive marketing.
Click Profit (March 2025).
Air AI (August 2025).
Rytr Consent Order Reopening (December 22, 2025). Order set aside; matter ongoing.

Data Broker Enforcement (CPPA)

Growbots Settlement (November 14, 2024). $35,400 fine, 177 days noncompliance with California Delete Act registration (February 1, 2024 - July 26, 2024).
UpLead Settlement (November 14, 2024). $34,400 fine, 172 days noncompliance (February 1, 2024 - July 21, 2024).
Accurate Append Settlement (July 29, 2025). $55,400 fine. Washington-based.

Academic and Research Papers

Anthropic Reward Hacking Generalization Research (November 2025). Establishes that reward-hacking generalization in production reinforcement learning produces alignment-faking, malicious cooperation, and sabotage. Cited in Module 1's tier 6 treatment and elsewhere.

Lee et al. "RLAIF vs. RLHF" (2023). arXiv:2309.00267. Foundational paper on Reinforcement Learning from AI Feedback. The book engages this paper in the RLAIF essay (Part I).

Li et al. "HRLAIF" (Tencent, March 13, 2024). arXiv:2403.08309. Tested RLAIF empirically and found 4.58% decrease in human satisfaction despite preference improvements. The "models get better at sounding right and worse at being right" finding.

Skalse et al. "Defining and Characterizing Reward Hacking" (2022). Provides the formal definition of reward hacking the trilogy uses.

Knight First Amendment Institute / Feng-McDonald-Zhang (July 28, 2025). "Levels of Autonomy for AI Agents." Five-level user-role framework. arXiv:2506.12469. Engaged in the framework comparison chapter.

TechPolicy.Press / Cooperative AI Foundation Multi-Agent Article 73 Critique (January 2026). Argues Article 73 as drafted assumes single-system, single-occurrence failures. Cited as forward-looking concern in Module 6.

Future of Privacy Forum Analysis of GDPR Article 22 Enforcement (2025). Documents how data protection authorities are examining "the entire organizational environment where automated decision-making is taking place."

University of Chicago Law Review. "The Law of AI is the Law of Risky Agents Without Intentions." Cited in the RLAIF essay on liability allocation when AI systems lack intentions.

Industry Surveys and Analysis

WRITER 2026 Enterprise AI Adoption Survey. Reports 97% of executives deployed AI agents in past year, only 12% reach production at scale. 36% lack any formal plan for supervising AI agents. 35% admit they couldn't immediately "pull the plug" on a rogue agent. 67% of executives believe their company has already suffered a data leak from unapproved AI tools.

Stanford Enterprise AI Playbook 2026. Analysis of 51 successful enterprise AI deployments found two-thirds had significant failed attempts before achieving value.

DigitalApplied 2026 Agent Production Analysis. 78% of enterprises have agent pilots, fewer than 15% reach production.

Stanford CodeX AI Adoption Research. 92% of AI vendors claim broad rights over user data, 17% commit to full regulatory compliance.

IBM 2026 CAIO Study. 76% of organizations have a CAIO in 2026, up from 26% in 2025.

EY 2025 AI Risk Survey. 99% of organizations report financial losses from AI risks, average $4.4M per company.

IBM 2025 Cost of a Data Breach Report. Shadow AI breaches cost $4.63M vs $3.96M for standard breaches.

Gartner Digital Governance Failure Analysis. 80% of digital organizations face governance failures absent structured ownership.

appliedAI EU AI Act Risk Classification Study. Analysis of 106 enterprise AI systems found 18% clearly high-risk, 40% with ambiguous classification.

Capgemini Research Institute. Rise of agentic AI: How trust is the key to human-AI collaboration.

Comparative AI Autonomy Frameworks

The seven frameworks engaged in the framework comparison chapter, in chronological order of publication:

NVIDIA Agentic Autonomy and Security (Harang and Sablotny, February 2025). Four levels (0-3) organized by determinism of execution paths. Audience: AI security architects.

Knight First Amendment Institute / Feng-McDonald-Zhang (July 2025). Five levels organized by user role. Proposes autonomy certificates as governance mechanism.

InfoQ Three-Tier Enterprise Architecture Framework (Natarajan and Ponnusamy, July 2025). Maturity-progression framework with Foundation, Workflow, and Autonomous tiers.

Tessl Five Levels of AI Agent Autonomy (Podjarny, September 2025). Coding-agent-specific. Explicitly mapped to SAE J3016 self-driving levels.

Vellum AI Six Levels of Agentic Behavior (December 2025). Capability-centric framework. L0 rule-based through L5 fully creative.

Cloud Security Alliance Levels of Autonomy (Reavis, January 2026). Six levels organized by granularity of human approval. Most directly comparable to this book's framework in count and SAE inspiration.

Swarmia Five Levels of AI Coding Agent Autonomy (March 2026). Domain-specific to AI coding agents.

Falconer Survey of AI Agent Autonomy Frameworks (Sean Falconer, November 2025). Comparative survey rather than new framework. Useful as landscape orientation.

Journalism

TechCrunch on 11x AI SDR (March 24, 2025). Investigation by reporting team documented 70-80% customer churn rate, $14M ARR vs. $3M reality after three-month opt-out, ZoomInfo and Airtable customer-claim issues.

Sifted on 11x AI SDR Culture (2025). Reporting documented former employee accounts of "barely functional" product and "constant fire-fighting."

Salesmotion on AI SDR Industry Reckoning (March 2025). Reporting on Reid Christian (CRV) "day of reckoning" comments and broader AI SDR industry trajectory.

TechCrunch on FTC Operation AI Comply (September 2024). Initial coverage of the five Operation AI Comply enforcement actions.

Computerworld on Microsoft 365 Copilot Testing (2024-2025). Multiple test reports documenting AI-generated content fabrications.

Vendor and Practitioner Publications

Legal Practitioner Analyses

Mayer Brown AI Procurement Contracting (February 2026). Treatment of vendor-customer contractual terms for AI systems. Engaged in Module 4 procurement chapter.

Latham & Watkins on Article 73 Draft Guidance (October 2025).

King & Spalding on US State AI Laws Plus Federal Preemption (January 2026).

Baker Botts US AI Law Update (January 2026).

White & Case on Executive Order 14365 (January 2026).

Brownstein on Colorado AI Act (March 2026).

Holland & Knight on GSA Proposed Clause (March 2026).

Ropes & Gray on White House March 20, 2026 National Policy Framework.

Clark Hill on Colorado-vs-Federal Preemption Tension (December 2025).

Industry / Practitioner Frameworks

Cloud Security Alliance Multi-Agent Risk Resources. OWASP Top 10 for Agentic Applications (December 2025). NIST AI Agent Standards Initiative RFI (January-February 2026). Microsoft Agent Governance Toolkit (April 2026).

RACI Framework Publications for AI Governance. Yields.io RACI matrix template, Arthur.ai governance framework content, Dataiku governance resources, Drata compliance materials. Engaged in Module 5 organizational design chapter.

Minimum Viable Governance Framework (Pundhir / askajay.ai, 2025). 90-day on-ramp framework adapted from NIST AI RMF for small organizations.

Reid Christian (CRV) Industry Commentary. General Partner at Charles River Ventures whose "day of reckoning" assessment of AI SDR space is engaged in the AISDR sequencing essay.

Book Authors and Concepts Referenced

Jim Reavis (Cloud Security Alliance). Author of the CSA autonomy levels framework engaged in the framework comparison chapter.

Guy Podjarny (Tessl, Snyk). Author of the Tessl five-level framework engaged in the framework comparison chapter.

Reid Hancock (Stanford). Subject of the January 2025 Minnesota federal court ruling on AI-fabricated expert declaration citations. Engaged in the CoPilot essay (Part I).

Derek Mobley. Plaintiff in Mobley v. Workday. The case structure he initiated has produced the leading AI hiring discrimination collective action.

Jake Moffatt. Plaintiff in Moffatt v. Air Canada. The case is the canonical Air Canada principle precedent.

About the Author

Hemendra Singh is co-founder and chief technology officer of YapperBot.com, an AI audio content platform, where his work spans product architecture, AI product development, and operational governance. He holds a USPTO patent-pending filing related to the platform.

His perspective draws on vendor-side architecture, buyer-side deployment, and executive accountability — the multiple vantage points operators hold simultaneously when governing agentic AI. As an architect at Bitt Inc., he architected central bank digital currency (CBDC) platforms — eNaira for the Central Bank of Nigeria and DCash for the Eastern Caribbean Central Bank — alongside an offline CBDC solution that was selected to win the 2022 G20 TechSprint Global CBDC Competition. He led the modernization of the New Jersey Transit fare collection platform. For federal government, he has consulted on modernization and digitalization of customer support functions for over a decade.

He holds a degree from the Indian Institute of Technology Delhi and has completed the Chief Technology Officer Program at Wharton and the Designing AI Products program at MIT. He can be reached at linkedin.com/in/hemendrasinghmd or hemen@yapperbot.com.

www.ingramcontent.com/pod-product-compliance
Lightning Source LLC
LaVergne TN
LVHW090355240926
848801LV00002B/49